D0079236

CONFLICT OVER THE WORLD'S RESOURCES

Recent Titles in
Contributions in Political Science
Series Editor, Bernard K. Johnpoll

Perforated Sovereignties and Internatioinal Relations: Trans-Soverign Contacts of Subnational Governments
Ivo D. Duchacek, Daniel Latouche, and Garth Stevenson, editors

New Approaches to International Mediation
C. R. Mitchell and K. Webb, editors

Israeli National Security Policy: Political Actors and Perspectives
Bernard Reich and Gershon R. Kieval, editors

More Than a Game: Sports and Politics
Martin Barry Vinokur

Pursuing a Just and Durable Peace: John Foster Dulles and International Organization
Anthony Clark Arend

Getting the Donkey Out of the Ditch: The Democratic Party in Search of Itself
Caroline Arden

Elite Cadres and Party Coalitions: Representing the Public in Party Politics
Denise L. Baer and David A. Bositis

The Pure Concept of Diplomacy
José Calvet de Magalhães

From Camelot to the Teflon President: Economics and Presidential Popularity Since 1960
David J. Lanoue

Puerto Rico's Statehood Movement
Edgardo Meléndez

Foreign Relations: Analysis of Its Anatomy
Elmer Plischke

CONFLICT OVER THE WORLD'S RESOURCES

Background, Trends, Case Studies, and Considerations for the Future

Robert Mandel

Contributions in Political Science
Number 225

GREENWOOD PRESS
New York • Westport, Connecticut • London

Library of Congress Cataloging-in-Publication Data

Mandel, Robert.
Conflict over the world's resources : background, trends, case studies, and considerations for the future / Robert Mandel.
p. cm.—(Contributions in political science, ISSN 0147–1066 ; no. 225)
Bibliography: p.
Includes index.
ISBN 0–313–26129–6 (lib. bdg. : alk. paper)
1. Natural resources. 2. Economic history—1971–
3. International economic relations. 4. Economic forecasting.
I. Title. II. Series.
HC59.M248 1988
333.7—dc19 88–15489

British Library Cataloguing in Publication Data is available.

Library of Congress Catalog Card Number: 88–15489
ISBN: 0–313–26129–6
ISSN: 0147–1066

First published in 1988

Greenwood Press, Inc.
88 Post Road West, Westport, Connecticut 06881

Printed in the United States of America

The paper used in this book complies with the Permanent Paper Standard issued by the National Information Standards Organization (Z39.48–1984).

10 9 8 7 6 5 4 3 2 1

Copyright Acknowledgment

Contents

Figures

Preface

The original seeds of thought for this book sprouted in the summer of 1978, while for the first time I prepared my course, "Politics of Global Resource Scarcity," which I teach at Lewis and Clark College. My enthusiasm for the project was at least partially spurred by disenchantment with existing literature, whose paucity of explicit theory on international resources issues in general and on resource conflict in particular had become painfully apparent. Long after my subsequent article on transnational resource conflict appeared in *International Studies Quarterly* in March 1980, I began developing a series of working drafts of this book, and only in 1987 finally completed work on the case studies. I wish to thank Lisa Burnett and Annette Kelley for their considerable help and support on the project. However, slow gestation and long time lags are no guarantee of an error-free manuscript, and any flaws are my responsibility.

Introduction

Throughout history nations have waged battles over resources. For example, some analysts claim that during World War II Italy, Germany, and Japan were motivated by the desire for land and natural resources, and even that in the Vietnam conflict the United States was motivated by the desire to protect its access to rubber, rice and tin.[1] While in a broad sense all wars involve an attempt to gain or consolidate control of resources, in recent years conflicts centering on natural resource issues have become especially prominent. Dahlberg and others assert that "war over natural resources continues to be a real possibility today," with nations ready to intervene if access to vital resources is jeopardized, and that the whole concept of national security has broadened to include these resources.[2] Pirages notes that resource conflict appears to be more plentiful as demand increases in the face of environmental constraints and disruptions and as the economic manipulation of scarce resources becomes a more crucial dimension of international war.[3] Finally, Quigg contends that environmental concerns seem to have heightened the tension between rich and poor nations and increased the range of international conflict issues.[4]

This book chooses global resource conflict as its focus because of (1) its increasing salience in international relations, (2) the dearth of extensive analytical treatments of the topic, and (3) the tendency of such clashes of interest to highlight the most critical environmental tensions in the international arena.[5] This study emphasizes the exploration of the causes and consequences of conflicts over the world's resources, along with the means for managing such conflict. The theoretical framework examines the relationship of resource scarcity, resource-related development decline, resource inequality, and resource interdependence with resource conflict. Five case studies, spanning the full range of resource conflict issues, serve to evaluate and refine this framework: (1) the global whaling confrontation occurring since 1972, (2) the Middle Eastern oil crisis of 1973–74, (3) the American grain coercion of 1975 and 1980–81, (4) the Soviet strategic minerals threat since 1980, and (5) the Chernobyl nuclear disaster of 1986.

Several characteristics distinguish this work from existing literature. First, the focus here is explicitly comparative across cases—identifying patterns in causes and consequences of resource conflict—unlike the prevailing analyses whose treatment of a single detailed case often withers in isolation. Second, this book emphasizes integration of concepts—synthesizing theoretical writings on conflict with those on a wide range of resource issues (endangered species, fossil fuels, nonfuel minerals, and pollution)—rather than restricting coverage only to narrow ecological concerns or to one resource issue. Finally, this study is one of the few to provide extended discussion of policy prescriptions for minimizing international resource conflict. It is indeed unfortunate that so much of the literature on resources and the environment has gained a reputation for being blatantly overemotional and nonanalytical, for insufficiently placing ecological problems in the context of political, economic, social, and military systems, and for largely ignoring articulated overarching theories; this book consciously attempts to avoid these pitfalls.

The book begins in Chapter 1 with a discussion delineating the characteristics of international resource conflict in the context of recent controversies over the nature of global resource scarcity. Chapter 2 presents theoretical propositions about the causes and consequences of international resource conflict in order to clarify prevailing assumptions. Chapter 3 explains the methodology for analyzing the case studies, and chapters 4 through 8 explore in some detail the five conflict cases. Chapter 9 draws general patterns about causes and consequences from the cases and juxtaposes these case findings to the theoretical contentions. Finally, Chapter 10 draws together policy prescriptions for minimizing international resource conflict.

1

The Nature of International Resource Conflict

This chapter delineates the book's domain of concern by evaluating perceived trends and presenting categorization schemes related to resource scarcity, international conflict in general, and resource conflict in particular. Because the current resource era has been so rapidly changing, no assumption is made that ongoing forms of scarcity and conflict will necessarily be those that dominate the future.

TRENDS IN GLOBAL RESOURCE SCARCITY

There has always been some relative scarcity of some resource—commonly defined as any natural substance relevant to human needs—for some group of people somewhere on earth. But in the 1970s the perception of scarcity grew. In 1972 the publication of *The Limits to Growth*, an apocalyptic forecast of the world running quickly out of needed resources due to diminishing reserves and a mushrooming population, provoked somewhat frenzied concern from both masses and elites in high-consumption Western societies.[1] The oil crisis of 1973–74 seemed an almost perfect fulfillment of the prophecies of doom, apparently demonstrating a truly global rather than local scarcity,[2] and spurred further acceptance of "a new dominant social paradigm" of scarcity.[3] It seemed not only that existing resource reserves were running low, but also that the costs of developing and discovering new resource deposits were becoming prohibitive because accessible areas—particularly in politically safe nations—appeared to have been already exploited.[4] However, trends in the 1980s appeared to call the alarmist resource attitudes of the previous decade into question: the emergence of the oil "glut" in the early 1980s has convinced some that there is nothing to worry about for the foreseeable future and that scarcities are local and temporary, and the publication in 1981 of *The Ultimate Resource* widely spread the view that resource scarcity was not occurring and in any case would not pose an insurmountable obstacle to growth.[5] Ecological pessimists find themselves increasingly on the defensive in debates with technological optimists.

Much of this disagreement appears to rest on differing notions about the definition of scarcity. Scarcity is an extremely complex and multifaceted concept, and many writers deal only with some of its facets or attempt to define it through rises in the cost of extracting resources or the price of buying them.[6] In this study, the scarcity of a given resource for a given nation at a given point in time is defined as the ratio of the human demand for the resource to the environment's ability to supply it. The demand is a function of (1) the size of the nation's human population; (2) the nature of the psychological addictions to resource consumption (which directly influences the resource use per person); (3) the presence of resource-intensive technology and/or significant economic purchasing power; and (4) the sociopolitical incentives for consumption (including its link to power). These four elements can dynamically interact, as, for example, a growing population can enhance the need for more resource-intensive technology in fueling skyrocketing demand. The supply is a function of (1) the physical depletion of the resource from the earth; (2) the perceptual awareness of the availability of the resource (which is often linked to the absence of relevant scientific knowledge); (3) the possession of technology and feasible economic means for extracting and processing the resource, in the context of geophysical obstacles to obtaining the resource; and (4) the political and social constraints on accessing the resource (such as protected wilderness areas). Again, an interactive effect is potentially present, exemplified by greater resource depletion frequently connecting with a lowering of political/social constraints on resource access and a greater disregard for environmental carrying capacities.

Having all of these dimensions of supply and demand may at first seem to overcomplicate and even confuse the basic definitions, but the added complexity seems necessary to deal with the subtleties of these two concepts and, most important, to show the variety of linkages to resource conflict. The threshold in the supply/demand ratio when scarcity commences is of course impossible to pinpoint. Over time, scarcity is frequently worsened, due to the increased difficulties of adjustment involved, when there is (1) a mismatch in the rates of change of supply and demand[7] or (2) supply and/or demand proves to be inelastic with respect to price/cost.[8]

As is readily apparent, the demand-based roots of scarcity are not completely independent of the supply-based roots. Nonetheless, there are demand-based and supply-based theories of resource scarcity: demand-based theories reflect a refusal or inability to adapt human demands to environmental supply; and supply-based theories reflect a refusal or inability to expand environmental supply to meet human needs.[9] Most analysts focus on the supply-based roots (and on supply-related solutions to scarcity),[10] but this emphasis may be questionable in terms of relative explanatory power, susceptibility of causes to manipulation, or value for revealing prescriptive cures.[11] The ecological/physical, psychological/perceptual, technological/economic, and political/social dimensions of scarcity are parallel for supply and demand and are central to this book's entire framework of analysis for international resource conflict.

Scarcity is commonly subdivided into "real" scarcity and "contrived" scarcity: real scarcity relates to the first dimension of supply and demand, the ecological/physical combination of a relatively depleted resource with a relatively large human population, while contrived scarcity relates to the last three dimensions of supply and demand, dealing with psychological/perceptual, technological/economic, and political/social conditions.[12] The prevailing assumption, which is certainly debatable, is that real scarcity is more severe and legitimate and less manipulable than contrived scarcity.

A related controversy concerns whether technological/economic conditions are facets of contrived scarcity, or, indeed, scarcity in general. Simon, for example, views human labor and human ingenuity and imagination as "the ultimate resource."[13] Furthermore, many have subscribed to the belief that "necessity is the mother of invention"—that technological and economic conditions will naturally adjust in an innovative and adaptive way to any resource supply shortages by (1) developing production techniques that rely on different resources, (2) expanding the discovery of new reserves of existing resources, or (3) increasing the efficiency of existing resource use. However, despite some historical evidence to support this assertion[14] and to show falling raw material prices and costs over time,[15] the problem of time lags (to be discussed later) can create a critical gap between supply shortage and demand adaptation, during which time scarcity occurs, and the inevitability of adaptation does not seem absolute. Thus technological/economic conditions or the absence of quick appropriate adaptation to cope with material shortages can associate with scarcity and would be contrived in the sense that these conditions are not a function of the "real" limits of the natural environment.

Figure 1 summarizes the notion of scarcity as presented in this study. Any of the four aspects of demand can be combined with any of the four aspects of supply to create scarcity. Scarcity can, of course, be mitigated by technological change, substitution of more abundant resources, trade, discovery of new resource reserves, recycling, and reducing consumption through a variety of means.[16]

Figure 2 displays the supply-demand trends that seem most likely to generate or reflect scarcity. Many economists assert that the price system would either prevent or correct such significantly disequilibrated relationships between supply and demand, and that resource supply often rises to match growing demands.[17] Doubtless it is true that the actual occurrence of such trends to the full extreme depicted has been relatively rare. But these trends appear to be especially applicable under two conditions: (1) when dealing with demand or supply that is inelastic with respect to price/cost, a condition frequently associating with high human dependence on a particular (nonsubstitutable) resource and high physical depletion of it; and/or (2) when dealing with situations where the market price of a resource bears little relation to the costs of extracting it (such as with oil). All three trends would have even more dire scarcity consequences if the rates of change escalated in opposite directions. The graph to the far right would appear to generate the most severe scarcity because the potential for adjustments conducive to a stable equilibrium seems smallest when the supply of a resource is decreasing at the same time the demand for it is increasing.

Figure 1
Dimensions of Resource Scarcity

	DEMAND									
REAL SCAR-CITY	*Ecological* (large population size)	* * *		* * *		* * *		* * *		* * *
	Psychological (addiction to consumption)	* * *		* * *		* * *		* * *		* * *
CON-TRIVED SCAR-CITY	*Technological* (resource-intensive technology)	* * *		* * *		* * *		* * *		* * *
	Political (political consumption incentives)	* * *		* * *		* * *		* * *		* * *
	SUPPLY		*Ecological* (physical resource depletion)		*Psychological* (unawareness of resource availability)		*Technological* (inability to access resource)		*Political* (access restric-tions)	
			REAL SCARCITY				CONTRIVED SCARCITY			

Figure 2
Time Perspective on Resource Supply-Demand Trends

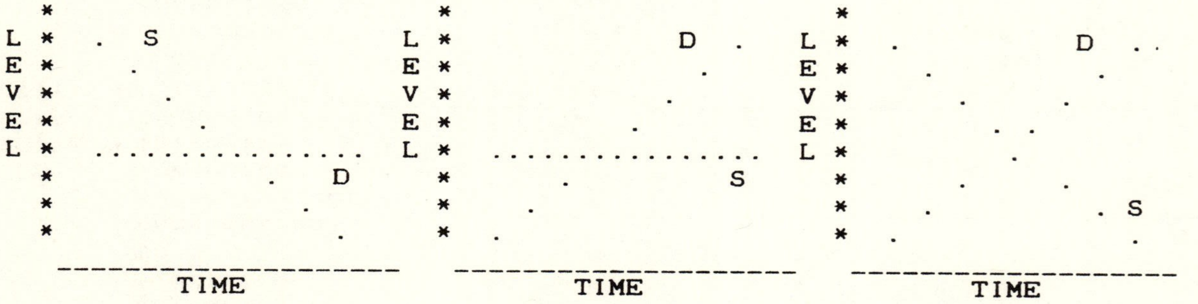

```
   *                        *                        *
L  *  . S                L  *          D .        L  *  .          D ..
E  *    .                E  *            .        E  *    .          .
V  *      .              V  *          .          V  *      .      .
E  *        .            E  *        .            E  *        . .
L  *  ................   L  *  ................   L  *          .
   *            .  D        *      .         S       *      .      .
   *              .         *    .                   *    .          . S
   *                .       *  .                     *  .              .
   --------------------     --------------------     --------------------
          TIME                     TIME                      TIME
```

Constant Demand and Decreasing Supply

Constant Supply and Increasing Demand

Increasing Demand and Decreasing Supply

TRENDS IN INTERNATIONAL CONFLICT

As with scarcity, views of the trends in international conflict have experienced fluctuation in recent years. In the latter half of the 1970s (after the Vietnam experience), some discerned a new emphasis on economic as opposed to military conflict.[18] This trend seemed to derive from the presumed obsolescence of territorial war as a viable means of achieving national objectives; the increased flexibility (in comparison to traditional military conflict) offered by economic instruments in a dispute; the reduced costs and risks—especially in terms of human life—of such economic instruments; and the reduced potential for entry or threat of nuclear weapons in an economic conflict. In contrast, in the first half of the 1980s, the world frequently has been characterized as facing a remilitarization of international relations. Shows of force and the use of physical coercion have proliferated as means of attaining goals, as confrontations in the Falklands, Iraq and Iran, Lebanon, Nicaragua, and Grenada illustrate. While identifying consistent conflict patterns across periods of time is a controversial process, the pendulum, at least for the moment, seems to have swung away from detente and the dominance of the economic North-South split back to the Cold War and the supremacy of the military East-West split.

Thus these fluctuating trends in perceived resource scarcity and international conflict are somewhat parallel and call into question any simplistic identification of linear monodirectional patterns. Both trends are consistent with the concept of cycles of integration and disintegration, isolationism and interventionism, and boom and bust.[19]

TRENDS IN INTERNATIONAL RESOURCE CONFLICT

Since the early 1970s there has been a largely unprecedented focus of global attention on resource conflict. Morgan colorfully explains the origins of this heightened concern:

> With the Russian "grain robbery" [in 1972] and the Arab oil embargo [in 1973–74], and price increases of both commodities, the old world of cheap oil and food gave way to unprecedented global inflation and shortages. America, which

> seemed so economically invulnerable, was revealed as just one more client of the oil kingdoms of the Middle East. Russia had to buy immense amounts of grain to carry out its plans. The superpowers looked wobbly. And less powerful countries—European nations and Japan, for example—were even more dependent on outside sources of raw materials and food. Politicians accustomed to viewing the world in terms of the ideological struggle of the previous two decades began to pay attention to the posted price of oil in the Persian Gulf, the spot price of Rhodesian chrome, and the corn "basis" at the Gulf of Mexico.[20]

Pfaltzgraff more specifically chronicles the ways in which natural resources have been viewed as critical during this period in exacerbating tensions within and across international alliances:

> Resource issues add to the vulnerability of some alliance members to resource blackmail, either by the Soviet Union or by producer states, or by both acting independently or in concert. Resource issues increase the potential for conflict between members of opposing alliances—for example, between Norway and the Soviet Union as a result of conflicting claims to oil and natural gas located in the Barents Sea, or between Japan and the Soviet Union over fisheries. Resource issues give rise to security problems, including the protection of offshore oil installations and the safeguarding of sea lines of communications vital to the transport of important resources. Resource issues have potentially far-reaching effects upon relations among alliance members resulting from competition for resources and disputes arising from claims to offshore resources. Examples include the dispute between Greece and Turkey over Aegean oil, and friction between the United States and Japan over the exploitation of fisheries.[21]

Many nations seem to have been unprepared to deal with the distinctive patterns of conflict generated by resource issues.

But not all observers have accepted as reality the nature and implications of the recent growth in conflict over the world's resources, as Kemp incisively explains. Pessimistic forecasts stress the increasing vulnerability of the industrialized nations' resource supplies to physical interruptions and price increases initiated by the Third World; the devastating effects of rapid population growth and high prices on the economic vulnerability of the poorest nations; the global scramble for offshore resources; and "the sporadic outbreak of actual fighting over resources in recent years" accompanied by increasing arms sales to states seeking to protect their resources and access routes. More optimistic forecasts retort, however, that economic interdependence is helping to reduce the potential for resource conflict; scarcity problems may be transitory; dire predictions of a chaotic global population explosion have not been fulfilled; internationally coordinated action, such as the United Nations Law of the Sea Conference, will eventually assist with ocean resources; and "military skirmishes to date have been more than offset by the rapid expansion of... cooperative ventures between a very disparate group of suppliers and consumers."[22] Nonetheless, even the most hopeful observers have accepted the increasing need to understand better and manage more decisively the causes and consequences of resource conflict.

Placing the trend of mushrooming resource tensions in the general context previously outlined of scarcity and conflict, a number of distinct forms of conflict over the world's resources have become prominent. While resource tensions can cross and have crossed the conceptual boundaries of these conflict types, the categorization scheme brings some coherence to the resource clashes of recent decades. The varieties identified seem compatible with international systems characterized by either severe or moderate scarcity and dominated by either East-West or North-South disputes.

One pattern indicates that powerful industrial states initiate conflict to achieve growth in resource consumption so as to satisfy their expanding resource appetites.[23] As developed states' population and technology grow, so do their resource demands, because much advanced technology requires greater quantities and wider ranges of resources; these states then turn outward because domestic sources cannot readily satisfy these demands. Frequently these expansionist desires are frustrated by embargoes, tariffs, and other restrictive regulations, along with the usual desire of target nations to maintain their own territorial sovereignty; the result, aided by the presence of large military capabilities, is conflict.[24] Russett claims that the world market is losing its force as a means of allocating supplies of natural resources and is being replaced by major powers seeking political/military means to attain secure access to vital raw materials.[25] Industrialized states appear to have particular problems in dealing with an energy shortage because energy has been relatively inexpensive and readily available for so long that they have become highly dependent on it. Conflicts of this type may be between the superpowers in either (1) offensive designs to gain control over the other superpower's imports of raw materials to permit resource denial or (2) defensive designs to safeguard one's own sources of raw materials from potential threats of denial stemming from the other superpower.[26] More specifically, this type of conflict may involve (1) military conflict to control, destroy, or protect a given resource; (2) military deployments to annex or protect a land or sea area containing or near potentially valuable resources; and (3) military strife over access routes to and from sources of supply.[27] The two resource-rich regions most hotly contested by the superpowers are the Persian Gulf and southern Africa. The struggles between the United States and the Soviet Union over grain and strategic minerals exemplify this form of conflict.

A second pattern has weaker states banding together and initiating conflict in order to achieve a redistribution of existing resources away from the strong states monopolizing them.[28] Contributing to this pattern is the increased insistence by Third World states that they be allowed to control the development of natural resources within their own boundaries.[29] Many developing nations have the potential power to withhold supplies or to use special pricing policies to attain this end, especially when (1) the target developed nations have an inelastic demand for a resource and no stockpiles of it; (2) no substitutes or alternative supply sources are available on short notice; and (3) the developing nations involved have large foreign exchange reserves. However, for many resources there is a real question whether developing nations have sufficient power to initiate such a battle, as "political influence does not automatically accrue through mere possession by a nation or group of nations of any industrial raw material."[30] The Arab oil embargo is a famous illustration of this kind of conflict.

Moving away from government-to-government conflict, a third pattern, which is more transnational in character and has not attracted as much notice, is the growing clash between private transnational lobbying groups oriented toward conservation and foreign governments violating these groups' environmental guidelines.[31] This type of resource conflict is based not on a desire to achieve growth in resource consumption or redistribution of resources, but rather on a desire to preserve the environment and thereby maintain the ecological quality of life. Confrontations by antinuclear environmental groups protesting the use of nuclear energy (particularly in the wake of the Chernobyl disaster) and by anti-whaling conservationist groups opposing Japanese and Soviet catches are typical of this form of conflict.

Aside from these three distinct forms of intergovernmental and transnational resource conflict, which constitute the focus of this book's coverage, intrastate resource conflict may occur: high-consumption resource consumers may battle against resource regulators in fights that reflect concerns over growth in resource consumption, redistribution of resources, or environmental preservation.[32] Pfaltzgraff notes that resource issues may increase "centrifugal" tendencies within states, especially if key resources are located near or in a region that has developed its own sense of separateness.[33] The enhancement of Scottish nationalism through the development of North Sea oil exemplifies this link to internal resource conflict, while the battle between environmentalists and industrialists over wilderness areas in the United States displays the more general pattern when subnational autonomy movements are not at stake. The relationship between intrastate and interstate resource conflict is subject to debate; it may be mutually supportive, or it may be hydraulic and zero-sum, as venting one's frustrations at one level may reduce the need or desire to do so at the other.

Despite the significant clashes of interest involved in all of these types of resource conflict, they seem relatively mild in the context of the broad spectrum of international conflict. There is general agreement that international resource conflict is likely to be nonviolent and fall short of all-out war.[34] For instance, Dryzek and Hunter describe the probable severity of clashes over environmental protection:

> Clearly, any lack of resolution of international environmental problems is unlikely to lead to large-scale violence or

> immediate system breakdown. . . . more likely than outright breakdown is progressive deterioration, or, at best, perpetuation of an unsatisfactory status quo—for example, gradual attrition in environmental quality.[35]

Furthermore, there is some evidence that resource issues are less likely to cause an existing dispute to escalate into all-out war than are other issues.[36] But Wallensteen points out that although economic resource weapons are not aimed at killing, they can indirectly threaten life through reducing the level of production, contributing to economic chaos, increasing production difficulties, or preventing the importation of a resource.[37] Moreover, Arbatov and Amirov voice the widespread concern that "in the present complicated international situation, any interstate conflict, even one rooted in the raw material problem, could develop into a conflict on a much larger scale."[38]

It appears important to recognize in conclusion that, although resource issues play a prominent role in modern international conflict, they do not invariably seem to dominate it. Kemp points out that the importance of resource concerns in recent wars has varied considerably and that sometimes these concerns play a secondary rather than primary role in creating tensions.[39] For example, an examination of the sixty-six major interstate border disputes between 1945 and 1974 found that only about a quarter of them had, as their primary focus, resource issues (rather than ethnic or political issues), while about a third of them involved resource concerns as at least a secondary element in the conflict.[40]

2

Theories about Causes and Consequences of Resource Conflict

This chapter explains the theoretical framework used for examining conflicts over the world's resources. The framework begins with the occurrence of a resource disruption and is followed sequentially by changes in (1) relative scarcity of the resource, (2) national development relating to resource productivity, (3) inequality of resource distribution, and (4) resource interdependence of the relevant parties. These changes in scarcity, development, inequality, and interdependence reveal the basic causal process involved in resource conflict, and then their impact on the initial resource disruption highlights the consequences of such conflict. Critical resource time lags occur across all of the stages identified and require separate discussion. Although the components of this framework are quite broad and inevitably not all-inclusive, there was a concerted effort to incorporate the most critical elements: transformation in a society's availability of raw materials (scarcity), productivity (development), relative status (inequality), and intertwining with other societies (interdependence). Figure 3 displays the components and linkages in this framework.

The theoretical strands contained herein, to be tested later through scrutinizing the case studies of resource conflict, identify tentative relationships based on a review of the often opposing assertions drawn primarily from the fields of ecology, economics, political science, and psychology. This analysis melds insights on resources with

Figure 3
Theoretical Framework of International Resource Conflict

RESOURCE DISRUPTION

\/ /\
\/ *Time Lag* /\
\/ /\

CHANGE IN RESOURCE AVAILABILITY:
Scarcity in
Supply-Demand Ratio

\/ /\
\/ *Time Lag* /\
\/ /\

TEMPORAL CHANGE IN DEVELOPMENT:
Decline in
Resource Productivity

\/ /\
\/ *Time Lag* /\
\/ /\

SPATIAL CHANGE IN EQUITY:
Inequality in
Resource Distribution

\/ /\
\/ *Time Lag* /\
\/ /\

CHANGE IN INTERNATIONAL LINKAGES:
Interdependence in
Resource Transaction

\/ /\
\/ *Time Lag* /\
\/ /\

RESOURCE CONFLICT

those on war and instability. Rather than simply presenting a propositional inventory, I intend to make preliminary judgments about linkages; the judgments are made using explicit logic to assess and qualify the sometimes vague and sweeping contending generalizations found in relevant literature. However, the level of generality remains quite high due to the inadequacies (particularly regarding consequences) of existing theory on the topic. Furthermore, studying tense aspects of human-environment interaction poses a particular challenge, because the influence relationships are often reversible and coextensive, making the isolation of cause-effect connections and the establishment of time sequences exceedingly difficult. To cope with the inevitable controversiality of some of its assertions, this study makes every attempt to articulate fully its reasoning and to refer to devil's advocate positions in the course of the analysis.

RESOURCE SCARCITY AND CONFLICT

Perhaps the logical starting point in an examination of the causes and consequences of resource conflict is scarcity itself. Resource scarcity may have catastrophic, transformational, or inconsequential impacts; may generate permanent or temporary losses of resources; and may entail global or local consequences.[1] Because there are many facets to the influence of scarcity on conflict, and they involve considerable controversy, this section first overviews the general stages of this impact and then more specifically considers when the broad assumptions made are weakest and strongest as a means of isolating the stage when scarcity is most tightly intertwined with resource conflict. Figure 4 summarizes the stages involved in this relationship.

The first sequence in the figure shows that regardless of whether scarcity is primarily supply-based or demand-based, it affects internal growth and development—in accordance with limits-to-growth logic—before external distribution and inequality because (1) scarcity's impact on distribution is indirect and only one component in a complicated system of consciously-formulated international exchange; and (2) scarcity's impact on growth is direct in terms of restricting available quantities and increasing access costs.[2] This reduction in growth may result not only from increasingly inadequate resources but also from environmental deterioration triggered by the pressures of scarcity. Decreased growth can both increase the de-

Figure 4
Impact of Scarcity on Resource Conflict

Reduced Access to Resources	Reduced Internal Growth and Development Rate (Limits-to-Growth Theory)	Reduced External Distribution of Resources (To Other Nations)	Reduced External Power and Influence (Geopolitics Theory)

PRE-SCARCITY SCARCITY HITS POST-SCARCITY

Expectations and Desires Escalate, Wants Viewed as Needs	Gap Occurs Between Expectations and Achievements	Gap Occurs Between Reputation and Achievements	Frustration, Relative Deprivation, and Rank Disequilibrium

Worsened Resource Demand/Supply Ratio	Increased Stress, Frustration, and Dependence	Increased Vulnerability to Disruption	Conflict Erupts (Worsening Demand/Supply Ratio)

mands for redistribution within a society and decrease the society's ability to distribute resources abroad.[3] The resulting stagnation tends to reduce the prevailing standard of living and quality of life as long as resource supply and demand remain out of equilibrium. The final impact is on a nation's external power and prestige—in accordance with geopolitical logic—and lags far behind the state's internal decay.[4]

After the onset of scarcity, the second sequence in Figure 4 indicates that two principal kinds of frustrating gaps can occur: the first relates to a nation's temporal ability to satisfy its own needs and desires (connecting to growth and development), and the second relates to a state's spatial ability to maintain respect in the eyes of others (connecting to distribution and power/influence). In the process, the time lags usually continue to grow between efforts to alleviate these gaps and desired results. Finally, two forms of resource frustration may emerge: relative deprivation, in which expectations from the past exceed present achievements, producing unhappiness about development conditions;[5] and rank disequilibrium, in which one's view of one's achievements exceeds one's reputation in the eyes of others, producing unhappiness about power recognition.[6] These two forms of frustration indicate that levels of scarcity—and, indeed, of development, inequality, and interdependence—need to be judged not according to absolute objective thresholds but rather "relative to what people want, feel, think and are told to want, feel and think" about resources.[7] In accordance with many theories of civil unrest, these two forms of frustration tend to be most acute if the existing scarcity reflects a dramatic change from a previous period of resource abundance.[8]

The final sequence in Figure 4 directly links scarcity to international resource conflict. Demand being above the level of equilibrium with supply can create stress, as this situation constrains a society's capacity to achieve progress—attaining preferred levels or rates of consumption/production.[9] A primary consequence of this resource stress on a nation is a fragile resource supply line, followed by virtually inevitable disruptions in supply, which—whether intended or unintended—are usually viewed as illegitimate and unacceptable. This condition of vulnerability can cause resource conflict in one of two ways: either the frustrated and vulnerable nation initiates conflict to reduce its vulnerability and gain access to more resources; or other nations see the vulnerability of the frustrated nation as an opportunity to initiate conflict when they have the advantage and their target is weak. Barnet notes that scarcity makes "resource-grabbing" a more powerful rationale for international coercive intervention,[10] and Ophuls claims that scarcity aggravates the competitive struggle within nations among groups attempting to gain economic benefits.[11] Wallensteen indicates that scarcity is the primary condition for turning resources into conflict-creating political weapons, along with the scarcity-related conditions of supply concentration in the hands of a few producers/sellers and demand dispersion among many consumers/buyers.[12] There appear to be overwhelming indications that scarcity increases confrontation and turbulence rather than cooperation or tranquility: nations have seemed to be far more likely to hoard resources during scarcity than to share them.[13]

Whether the conflict resulting from resource scarcity reaches actual military confrontation may in the end depend on the level of coercive capabilities available to affected parties.[14] For example, scarcity of strategic minerals or other militarily vital nonrenewable resources could limit the scope or likelihood of war or could force a zero-sum tradeoff (due to high resource prices) between national expenditures on armaments and those on economic development.[15] However, Kemp finds that resource scarcity generally amplifies concerns for national security and willingness to escalate military expenditures (as has been recently evidenced in the Persian Gulf):

> Turning to the impact of resource scarcity on military planning, it is already clear that the combined impact of the energy crisis and the search for alternative resources has begun to influence the configuration and mission of peacetime military forces. Increasing numbers of countries are buying and deploying forces either to deter resource conflicts or to make sure they can contain them if they occur.[16]

Furthermore, attempts to discover and develop new resources to cope with scarcity may create significant military spinoffs, such as in the cases of ocean and outer-space exploration.[17]

Despite the convincing nature of these general arguments about scarcity retarding internal growth and external distribution while promoting frustration and conflict, they involve numerous logical weaknesses that serve to help to limit the applicability of the theoretical claims. Barnett and Morse summarize well the view that scarcity may have effects that society would deem beneficial rather than detrimental:

> Thus, the increasing scarcity of particular resources fosters discovery or development of alternative resources, not only equal in economic quality but often superior to those replaced. Few components of the earth's crust, including farm land, are so specific as to defy economic replacement, or so resistant to technological advance as to be incapable of eventually yielding extractive products at constant or declining cost.[18]

Substitution, new technologies, new trade routes, and changing political alignments are all methods to escape the impact of short-term local scarcity, whether real or contrived; and societies may accept certain manifestations of scarcity and learn to live with higher prices or lower quantities without heightened tensions.[19] The global stock of resources is not fixed, and the incessant and unpredictable change in which resources are scarcest at any point in time continues to alter nations' perceptions of scarcity.[20] But considerable ingenuity is required to create these changing adaptive responses to scarcity at a pace which keeps up with the growth of scarcity itself: Barnett and Morse therefore qualify their own assertion, showing by implication when the logic behind the stages of scarcity's impact shown in Figure 4 are weakest, through contending that constructive adaptation to scarcity is most likely when production processes are complex and decisions are guided by economic rationality, as is most common in advanced industrial societies rather than developing ones.[21]

Isolating the stage when scarcity has its greatest impact on resource conflict necessitates examining its level and rate of change. While extreme abundance itself contains the seeds for certain kinds of frustration and conflict, it appears that the more a society moves from moderate abundance to severe scarcity, especially through the trend of escalating inelastic demand combined with plummeting inelastic supply previously suggested in Figure 2, then the more conditions deteriorate and the greater the potential for resource conflict. Hveem more concretely delineates this notion of severe scarcity, which is most likely to promote resource conflict, as involving (1) a military and industrial system highly dependent on scarce resources (inelastic demand); (2) monopolistic or near-monopolistic control of these needed scarce resources, particularly by nations across political or ideological divisions (inelastic supply and contrived scarcity); (3) significant natural physical limitations on resource supply, particularly when accompanied by a major "supply imbalance" (inelastic supply and real scarcity); and (4) a tense international context regarding the scarce resources, especially involving a struggle for independence or disputed territory.[22] While scarcity at some levels can certainly be an asset to some societies, and is even considered by some to be a positive virtue, severe scarcity would seem only to frustrate societies bent on rapidly expanding internal development and external influence.[23] Moreover, when the level of scarcity worsens suddenly and massively (and surprisingly) rather than gradually and incrementally, the chances of resource conflict seem greatest because of the lack of time available for human expectations and social structures to make or accept needed adjustments.[24] According to theories of destabilizing change, the presence of rapid change produces conflict due to the disorientation and disequilibrium produced: the changes in various aspects of a society are unlikely to be coordinated or to occur at the same rate, and so upheaval results as new hierarchies develop.[25] Furthermore, there is reason to believe that when scarcity is perceived as the product of intentional human manipulation (as with contrived scarcity) rather than unintended and/or natural forces (as with real scarcity), the resentment is greater and the conflict potential is higher.[26] The conflict cases to be considered help to sort through this tangled web of conditions when resource scarcity produces the greatest tensions.

Once conflict does erupt, the consequences seem to worsen scarcity further by generating more disruptive change and by exacerbating existing environmental problems.[27] Stein and Russett point out that during conflict the extraction and utilization of resources can increase more than during peacetime,[28] and Galtung shows that the military system in particular is a major consumer of fossil fuels and scarce mineral resources, as well as of land and water during maneuvers.[29]

DEVELOPMENT AND CONFLICT

As indicated earlier, scarcity has its most direct and immediate effects on nations' internal development, a temporal issue reflecting productivity within societies. Despite a swirling debate on this issue, scarcity in general seems to have its greatest impact on the least fortunate nations (and individuals) in the world.[30] Furthermore, increasing scarcity tends to increase the proportion of the world that experiences poverty and deprivation without at the same time lowering expectations of material well-being. The claim can certainly be made that developed nations, despite their resource abundance, have a more fragile societal and industrial infrastructure, vulnerable especially to disruption in the supply of nonrenewable resources (energy); but what is at stake here is not survival of life itself in these states so much as the survival of a particular way of life, and their capacity for some adaptation seems reasonably high due to their technology, internal resource reserves, and interdependencies with each other.[31] Developing nations, on the other hand, have lower aggregate resource needs, but they have little resiliency and small shortages—especially in renewables (food)—can really hurt the chances even of bare survival. Although developed nations hope for internal growth and developing nations hope for external redistribution, developing states face the most severe tradeoffs between growth and distribution (both internally and externally), as well as among the resource goals of economic efficiency, environmental protection, and political security. Severe scarcity seems to cause nations which are on the verge of achieving development objectives to fail in that effort. The price rises resulting from such scarcity are often simply too much to bear for capital-poor nations, even with the presumed availability of foreign loans. The inherent weakness of most Third World currencies, along with the extra impoverishment many of these states have recently experienced due to the debt crisis, has certainly exacerbated this situation. Furthermore, resiliency to resource supply shortages and collapses is much lower for the developing nations.[32] Paradoxically, because of the severity of these internal and external economic problems, developing nations treat environmental issues in general as being a much lower priority than do developed nations.

These general links between scarcity and development need qualification by type of resource and type of scarcity. When one is focusing on nonrenewable resources, especially the nonfuel minerals, the impact of scarcity is clearly most dire in industrialized nations. Similarly, while real scarcity may hit hardest in developing nations, which often lack the capacity for creative adaptation to raw materials shortages, contrived scarcity—particularly that triggered by political manipulation or revolution—frequently causes the greatest uproar in developed nations because they are so unused to such deprivation; the developed world has a long-accepted tradition of imposing contrived scarcity on the less fortunate nations, reinforced by the legacy of the colonial era, but is not used to being the victim rather than the perpetrator of this form of resource shortage. On the whole, however, the general claim that the industrialized world has not had its development significantly hampered by resource scarcity stands up to historical scrutiny.[33]

It is indeed ironic that, although developing nations are the hardest hit by resource scarcity, frequent claims emerge that their international influence has increased due to this scarcity. Arad and Arad assert that the balance of power may be changing in favor of the Third World,[34] and Pfaltzgraff contends that these states now possess "leverage of unprecedented dimensions in relationships with industrialized, resource-consuming countries."[35] Szuprowicz concludes that after the Arab oil embargo many developing nations supplying critical resources to the West realized that they had the potential to exert political and economic blackmail, particularly against Western Europe and Japan, which he characterizes as the resource "have-nots" of the world.[36] But these warnings—which imply that through their scarcity-induced leverage Third World nations might be able to make progress in their development goals, while parts at least of the West stagnate—seem to understate the resiliency of the developed states in responding to such leverage attempts and to overstate the resiliency of the developing world in implementing and sustaining such leverage.

Given the previously discussed importance of subjective resource thresholds, the relationship between development and resource conflict requires consideration of societal needs, expectations, or wants/desires. The potential for resource conflict does not seem high within or among many of the least developed (Fourth World), subsistence-level, need-based nations, because such societies, although hardest hit by scarcity, do not have high or rising expectations about fulfillment of basic resource needs. Moreover, these unfortunate nations rarely possess adequate coercive potential to initiate resource conflict.[37] However, Third World nations with high development goals do seem more conflict-prone (despite their limited resource leverage), because such states may experience rising expectations and competitiveness combined with an inability to satisfy these expectations and unacceptable time lags.[38] Many such states, due to rapid population growth, perceive industrialization not just as a desire but as an urgent need.[39] Furthermore, deteriorating conditions in the Third World pose threats to national and international security, thus encouraging military escalation.[40] For such developing nations with expectation-achievement gaps, "raw materials are the major, and often the only, means available to these countries to engage in this conflict."[41] Similarly, advanced societies appear conflict-prone because their expansive focus on lavish wants/desires (as well as expectations) leads to virtually insatiable resource appetites. Thus those nations hardest hit by resource scarcity because of their development level are not necessarily those exhibiting the greatest potential for resource conflict.

In light of the preceding discussion, resource conflict seems most likely when a nation experiences a development decline reflecting a growing gap between its expectations of natural resource productivity and its actual achieved levels. Figure 5 displays the three patterns of growing expectation-achievement gaps. The graph to the far right would appear to be the most conflict-prone when dealing with resource issues because of the existence of the greatest possible shock to the society—just when development expectations are rising, development achievements are falling. These expectation-achievement gaps prove to be particularly important in consideration of the specific conflict cases, where traditionally the psychological component—national resource expectations—is often overlooked at the expense of the more tangible national resource achievements/productivity.

The consequences of the occurrence of resource conflict back on national development are ripe with controversy. Stein and Russett indicate that war both promotes and retards growth: war can kill people and destroy productive facilities but may also increase demand and overall national productivity.[42] Galtung argues that military activity generally has a negative impact on development goals.[43] Given that resource conflict does not generally involve violence or large-scale military activity, it seems likely that such conflict would not inevitably alter national development in the aggregate but rather promote growth in some sectors and decline in others.

RESOURCE INEQUALITY AND CONFLICT

A scarcity-induced transformation in a nation's development seems highly likely to change that nation's capacity to distribute resources abroad, affecting the spatial issue of inequality with other nations. The preceding discussion's general conclusion that severe scarcity hurts development more in poor nations than in rich ones would logically seem to lead to overall increased inequality in the international arena. Greater scarcity usually does appear likely to produce greater inequality, until really extreme scarcity occurs, as long as there is some freedom in consumption/production patterns.[44] Midlarsky indicates that, particularly when the increased scarcity is a function of increased demand in the form of population growth, the consequences are severe inequality and zero-sum competition.[45] While some argue idealistically that greater scarcity could promote sharing due to the recognition that "we're all in the same boat," and that equitable resource exchange could promote everyone's well-being in the long run, such a pattern would seem particularly unlikely when dealing with a society which is hit by scarcity and has an accepted tradition of at least some past inequality and private ownership of resources. At the highest levels of scarcity, of course, all would be brought down to the subsistence level (or below).

Contrived scarcity seems to promote the greatest maldistribution, because the poorest do not have access to technology or economic means for gaining resources and are often under severe political and social constraints (only some of which are self-imposed). The resulting maldistribution appears to be greatest in nonrenewable resources, because (as compared to renewables) a far more restricted set of nations are blessed with such raw materials within their borders and a far smaller group of states would be able to afford such resources at the scarcity-induced higher prices.

Moreover, the presence of inequality in resource distribution appears to generate frustration under scarcity. Perfect equality could certainly generate some frustration because select individuals, groups, or nations might feel that they deserve to be superior. Nevertheless, the greatest frustration apparently derives from severely unequal resource distribution, particularly if the inequality represented a dramatic shift from previous conditions, because of the feelings of

Figure 6
Impact of Inequality on Resource Conflict (Under Scarcity)

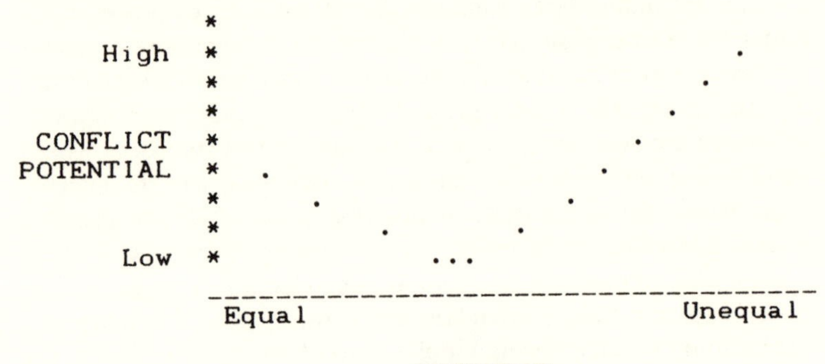

deprivation and rising expectations and the seeming incompatibility of scarcity and justice.[46]

The threads of this discussion of frustration lead to some intriguing conclusions about inequality and resource conflict shown in Figure 6. Complete equality might produce conflict due to the lack of responsiveness to and recognition of differing levels of talent or productivity; but—in accordance with general inequality-conflict theories going back to Aristotle, Karl Marx, and Alexis de Tocqueville—severe inequality seems to create the greatest chances of conflict. Resource inequality in particular seems to promote conflict because the lack of a safety net at the bottom and the intolerable discrepancies in resource consumption and living standards may entice nations which lack needed resources to become desperate.[47] Inequality may cause international conflict through (1) the increasing resource demands of the states on the top of the pecking order, which have the most to lose but at the same time the most power to hold on to what they have,[48] engaging in a kind of "resource imperialism" against other states;[49] or (2) the belligerence of resource-rich states lower on the pecking order—"the uneven concentration of resources in a relatively few countries, along with sharp differences in ideology, living standards, income levels, and forms of government, creates a fertile ground for international instability."[50] Internal conflict may result especially when there is inequality in land distribution.[51] Scarcity-induced inequality within a nation can lead to external venting of frustrations through a desire to identify and punish external scapegoats, in much the same manner as scarcity-induced inequality across nations. Existing inequality would seem to heighten most the potential for resource conflict when a society was not used to inequality, rejected steady-state values such as frugality and communalism, and/or placed tremendous value on its unequal dimensions.[52] As the conflict cases will illuminate, the size of ongoing resource inequality is not nearly as important as its rate of change and the way in which a society reacts to it.

As to the impact of resource conflict back on international inequality, there remains little doubt that war can play a major role in determining the distribution of power among states.[53] Galtung suggests that military activity reduces international equity,[54] but Stein finds that within nations war may decrease inequality in the distribution of goods.[55] According to these fragmented threads of general theory, it would appear that, with particular regard to resource conflict, clashes do not equalize international status and power but instead cause those states in the most secure resource positions to draw still further away from those in the least secure positions. However, given the rapid ebb and flow of changes in resource demand and supply, any moves upward on the international pecking order would probably be temporary.

RESOURCE INTERDEPENDENCE AND CONFLICT

The changes in resource inequality produced by severe scarcity naturally connect with alterations in the resource interdependence among nations. Earlier in the chapter, Figure 4 showed the central importance of interdependence and vulnerability in the sequence of stages linking scarcity to resource conflict. Though a debate exists (related to the preceding sharing/hoarding discussion) about whether scarcity promotes interdependence or divisiveness, scarcity does seem to enhance the need for interdependence because of the increased efficiency (often through economies of scale and comparative advantage) in resource use introduced by sharing.[56] Friedland and others note that scarcity-induced resource price rises have brought many nations "to a characteristically distasteful recognition of just how much they depend on each other's economic actions."[57] Inequality

Figure 7
Impact of Interdependence on Resource Conflict (Under Scarcity)

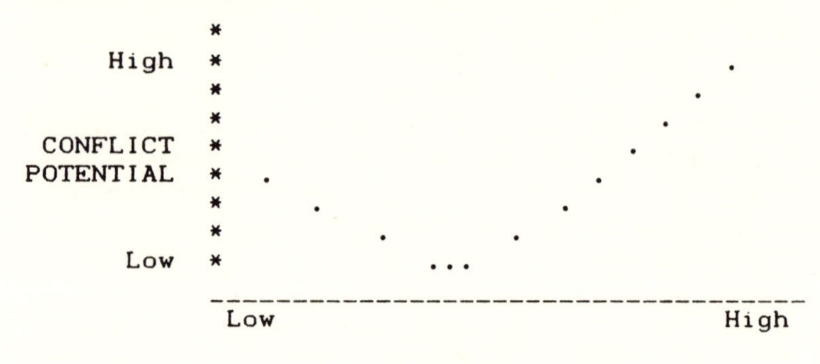

in resource distribution seems likely to increase the chances for skewed interdependence, particularly when this inequality is the product of contrived rather than real scarcity, because of the differential bargaining power possessed by nations with respect to their disparate levels of ecological, technological, and political assets.

A complex relationship exists between severe scarcity and resentment of resulting resource interdependence. On the one hand, interdependence in nonrenewable resources may seem more likely to be forced and resented because (unlike renewables) nations not endowed with such resources have a distinctly limited supply. On the other hand, however, interdependence in renewable resources could generate more resentment because many renewables such as food are widely viewed as basic human needs (unlike nonrenewables), and nations without adequate food for their population may view it as illegitimate that nations with food surpluses only send out a fraction of their extra stock and/or demand much in exchange. So severe scarcity of renewables and nonrenewables generates different kinds of resentment of resource interdependence.

Figure 7 displays the general relationship between resource interdependence and resource conflict. High levels of interdependence may often lead to conflict through the discomfort and perceived vulnerability of not having a secure internal source of vital resource needs and through the continuous competition and grating negotiations over renewed access to these resources.[58] Generally speaking, the higher the material standard of living, the greater the external dependence, and the greater the aggregate vulnerability to disruption.[59]

But there is considerable debate about the relationship between interdependence, vulnerability, and conflict. One common view is that, "with respect to natural resource supplies, dependency *is* vulnerability."[60] Blechman implies that the dangers of foreign resource dependence are greatest when the resource is not available in adequate quantities, the market price is not reasonable, the degree of dependence is high, few alternative suppliers are available, domestic deposits are low, instability or antagonism exists in the source countries, and little chance exists of using substitution of alternative resources or conservation measures.[61] Pfaltzgraff concurs with this pessimistic view of resource interdependence:

> Resource issues provide evidence of the inherent limits of globalism, regionalism, and international organizations in the resolution of "global" issues in an "interdependent" international system. Instead, we have witnessed a growing "politicization" of international economic relations at the state-to-state level and increased potential for conflict as a result of resource issues.[62]

However, others feel that interdependence does not automatically connote vulnerability and need not be viewed as intrinsically conflict-producing, at least in part because they see the dangerous combination of conditions cited earlier as exceedingly rare.[63] Moreover, Maull argues that resource interdependence may constrain conflict and reduce vulnerability, because people are aware of the increased potential for backfire effects of the use of resource weapons in such an interconnected world, a concern particularly prominent in the case of the oil crisis.[64]

Intense interdependence relationships seem particularly likely to lead to resource conflict if they reflect significant increases over past interdependence levels.[65] Nations may find if difficult, even under high interdependence, to take effective joint action in order to avert a conflict,[66] at least partially because of of the deep-seated "tribalism" in the international system.[67] On the other hand, extremely low levels of interdependence, reflecting considerable independent self-sufficiency, may promote some conflict—in the absence of total isolation—due to the lack of restraint imposed by international ties, needs, and obligations.

Skewed interdependence (or dependence) seems much more likely than reciprocal interdependence to generate resource conflict because (1) one nation is probably taking a lot more than it is giving, fostering resentment, frustration, and perceptions of illegitimacy in the other nation; and (2) the first nation might become increasingly unreasonable in its demands. Even if the resource exchange between interdependent developed and developing nations is not as skewed or exploitive as is commonly claimed, the key is that developing nations perceive exploitation and resent it.[68] In those instances where interdependence is neither skewed nor perceived as skewed, the potential for resource conflict may be minimal, especially if there exists a stable, legitimate, and mutually satisfying exchange relationship. In general, resource interdependence seems especially likely to raise the potential for resource conflict when the nations involved do not understand each other well, do not communicate with each other well, and/or are traditional antagonists.

The consequences of resource conflict appear usually to be to perpetuate or increase international interdependence. In many cases, a conflict-related disruption in nations' foreign supply of imported raw materials may stimulate attempts by affected states to reduce this dependence and move toward self-sufficiency; but the ability to achieve this end has been limited for many states, and in the recent era the international system does not appear structurally to permit this kind of withdrawal on a widespread basis.[69] Galtung claims that in today's international system military activity implies dependence rather than self-reliance;[70] and Snyder and Diesing assert that ongoing conflicts can create interdependence among allies and adversaries, due to dependence on allies' goodwill in settling conflicts and on enemies' capacity to inflict harm during conflicts.[71]

ROLE OF RESOURCE TIME LAGS

Time lags provide both opportunities and obstacles with regard to global resource tensions; they are present within and among the occurrences of resource disruptions and transformations in scarcity, development, inequality, and interdependence. Time is in some senses our scarcest resource, and thus is at the heart of international environmental problems.[72] Although these time lags have received occasional attention, they are the most underscrutinized element of the global resource picture.[73]

Figure 8 displays the principal different kinds of time lags faced in resource issues. As can be seen, a continuing cycle of ecological, psychological, technological, and political time lags (parallel to the dimensions of scarcity discussed earlier) is present in dealing with the environment. The first step involves human interaction with the environment, which occurs continually and may be intentional or unintentional. Next is the effect of this interaction on the environment, usually disruptive in some way to existing ecosystems. After some time, humans notice the ecological changes that have resulted from their behavior, then humans frequently attempt to develop corrective responses in order to counteract some of the disruptive effects of their behavior. Finally the cycle begins again as these corrective responses represent a new form of human interaction with the environment. The desire to extract and consume natural resources, which is the starting point of all resource conflict tensions, is also commonly the starting point of this cycle of time lags and of human disruption of the environment. But Berry and Johnson explain that "environmental changes due to human activity are seldom wanton acts of needless destruction" and instead "are the inadvertent, counterintuitive result of efforts to use resources."[74]

Severe scarcity has a dramatic impact on changing the length of time lags. In general, scarcity seems to make ecological (natural) time lags shorter (speeding up the possibility of environmental damage), and psychological, technological, and political (human) time lags longer, slowing repair of human-initiated environmental damage. This last impact is particularly problematic because of the short and incrementalist planning horizons of most policymaking institutions.[75] Real scarcity has as its greatest effect shortening the ecological time lags, because it makes the ecosystem more vulnerable to damage; while contrived scarcity has as its greatest effect lengthening the psychological, technological, and political time lags, because it constrains human ability to monitor the environment and to formulate and implement corrective responses through the political system. One could certainly argue that scarcity might shorten rather than lengthen technological and political time lags, assuming there were universal recognition that scarcity existed and demanded immediate action, but such intense disagreement could emerge about the exact nature of resource problems and solutions that the involvement of more concerned parties might serve only to complicate the process.

The impact of time lags on resource conflict (under severe scarcity) follows naturally from the preceding analysis. Shortened ecological time lags, which hasten the emergence of environmental decay, increase conflict-producing tensions because they have a negative impact on the supply side of the resource supply/demand ratio. Lengthened psychological, technological, and political time lags, which reduce human ability to take corrective action, enhance the chances of resource conflict, because they increase the tension-producing scope of the environmental disequilibrium. These effects of time lags on resource conflict serve to amplify the previously discussed effects of scarcity, development, inequality, and interdependence on conflict; they prove to have pervasive influence in the resource conflict cases discussed in this study. Without quick corrective responses, human-initiated environmental degradation naturally grows over time and becomes less restorable. The consequences of resource conflict on time lags can thus be to worsen them and further inhibit their ability to deal with environmental degradation and resource scarcity. The relevance of time lags to resource conflict is often greatest when there is only partial understanding about the length and means of altering these lags, leading to escalating fears and frustrations. However, time lags could limit resource conflict in at least one regard: lengthy transportation time lags could permit vulnerable consumer nations to continue to receive vital imported resources for weeks after a supply cutoff was announced.[76]

SYNTHESIS

The preceding theoretical analysis reveals a number of specific conditions most likely to generate resource conflicts. Severe resource scarcity is most associated with these clashes when it is unexpected, rapid, massive, and intentionally contrived, especially when it reflects increasing inelastic demand and decreasing inelastic supply. Development concerns most directly connect with conflict when a nation experiences economic decline combined with a growing gap between its expectations and achievements relating to resource productivity,

3

Methodology for Case Study Analysis

This book uses the "focused comparison" case study approach as the means of examining the causes and consequences of conflicts over the world's resources. While casually used for decades, this focused comparison method was only recently formalized and systematized by George and Smoke.[1] This technique examines multiple cases and draws its conclusions by making comparisons among them with respect to a prespecified set of investigative areas. Differences as well as similarities emerge from this kind of analysis, with the result being conditioned generalizations about patterns. The approach sacrifices the high reliability of statistical analysis of aggregate data for greater policy relevance and more comprehensive coverage of hard-to-measure variables and relationships.

This last advantage is particularly useful for the study of resource conflict. As the previous chapter's theoretical discussion indicates, this phenomenon involves considerable controversy in definition, wide scope in elements covered, and tricky cause-effect connections reflecting synergistic, symbiotic linkages. These qualities appear to inhibit quantification, and the assessments needed for statistical evaluation seem to invalidate much of the data, because they rely on untested

Much of this chapter is parallel to (and cites from) the methodology chapter in my previous work, *Irrationality in International Confrontation* (Westport, Conn.: Greenwood Press, 1987).

assumptions and leaps of logic. At the same time, empirically exploring resource conflict simply through a single case study, as is too often the case in existing literature, or through anecdotal references to past incidents would provide little basis for sound generalization or meaningful policy prescription.

BASIS FOR CASE SELECTION

The sample of five case studies selected for analysis is designed to be representative of recent conflicts over the world's resources. There has been considerable discussion about the last quarter of the twentieth century being a new era of "eco-politics" involving heightened awareness of resource issues; many consider this period to have begun with the United Nations Conference on the Human Environment in 1972.[2] All of the cases included in this study occurred after that date (with two, concerning whaling and strategic minerals, still ongoing at the time of this writing in the summer of 1987), providing a reasonably homogeneous international context as a backdrop. Analyzing conflicts that occurred so recently reduces the availability of broad and detached historical perspective, but at the same time increases the relevance of the evaluation to current policymaking.

The cases included all incorporate significant clashes of interest involving direct concern and policy adjustment by contending national governments and broad appeal to the international community as a whole. These bilateral transboundary resource disputes, such as those that frequently occur among nations over water rights, are too narrowly focused for this study; these are excluded along with minor international resource tensions, ecological controversies that have not yet gained prominence in the political arena, and transnational resource strife between private groups, not national governments.

Each case is most directly representative of a different type of resource issue that has been the focus of concern recently in world politics: endangered species, fossil fuels, food, nonfuel minerals, and pollution. The cases also cover the three forms of international resource conflict discussed in Chapter 1: powerful states trying to achieve growth in resource consumption, weaker states trying to achieve resource redistribution, and transnational conservationist groups trying to achieve environmental safeguards. Finally, the conflicts included revolve around values for both security of resource access and preservation of the natural environment.

The cases geographically span many parts of the world, but they more heavily involve developed rather than developing nations, because the totality of international resource conflicts for the last decade and a half has reflected Third World initiation considerably less than many analysts had predicted. The selection process consciously attempted to avoid an ethnocentric focus exclusively on cases involving the United States.

Each case identifies initiators and victims (the focus is largely on external victims), not to assign guilt to nations responsible for launching conflict and sympathy for those who suffered its impact, but rather to clarify the beginning and direction of the conflicts in order to evaluate their causes and consequences. Indeed, initiators identified may not have commenced conflict intentionally or through premeditated planning, and victims identified are only the principal nations affected by conflict.

Figure 9 displays the background of the cases, and Figure 10 places the cases into conflict categories. The final category depicted is the severity of resource conflict, an admittedly controversial judgment for which the case analyses provide substantiation. The criteria for evaluating severity are, in order of priority, (1) the importance of the resource issue to the national interests of the parties to the conflict; (2) the extent to which these parties contemplated military intervention to deal with the resource issue; (3) the degree of unwillingness by these parties to compromise or reach a conflict settlement; and (4) the frequency and explicitness of antagonistic actions or statements by these parties.

CAUSES AND CONSEQUENCES IN THE CASES

The focused comparison of the cases encompasses a relatively brief description of the course of each resource conflict, followed by detailed analyses of its causes and consequences. Referring back to the theoretical framework of Chapter 2, the causes encompass both the origins of the initial resource disruption and the roots of the clashes that occurred in response to this resource disruption; thus causation may be a two-stage process involving both short-term triggers and long-term precipitants. As to consequences, this study restricts its

Figure 9
Background of Resource Conflict Cases

	WHALING CONFRONTATION	*OIL CRISIS*	*GRAIN COERCION*	*STRATEGIC MINERALS THREAT*	*CHERNOBYL NUCLEAR DISASTER*
DATES	1972–Present	1973–1974	1975, 1980–81	1980–Present	1986
INITIATOR	International Whaling Commission, Transnational Eco-Groups	Organization of Arab Petroleum Exporting Countries	United States	Soviet Union	Soviet Union
VICTIM	Japan, Soviet Union	Japan, United States, Western Europe	Soviet Union	United States	Western Europe

Figure 10
Typology of Resource Conflict Cases

	WHALING CONFRONTATION	*OIL CRISIS*	*GRAIN COERCION*	*STRATEGIC MINERALS THREAT*	*CHERNOBYL NUCLEAR DISASTER*
TYPE OF RESOURCE ISSUE	Endangered Species (renewable)	Fossil Fuels (non-renewable)	Food (renewable)	Nonfuel Minerals (non-renewable)	Pollution (non-renewable)
TYPE OF RESOURCE CLASH	Strong States and Eco-Groups Fight Strong States to Achieve Resource Safeguards	Weaker States Fight Strong States to Achieve Resource Re-distribution	Strong States Fight Among Themselves to Achieve Resource Growth	Strong States Fight Among Themselves to Achieve Resource Growth	Strong States Fight Among Themselves to Achieve Resource Safeguards
TYPE OF CONFLICT VALUE	Environmental Preservation	Resource Security	Resource Security	Resource Security	Environmental Preservation
SEVERITY OF CONFLICT	Low	High	Medium	Low	Medium

consideration to short-run effects of resource conflict—usually within one year after a resource disruption—in order to concentrate on those impacts that most directly result from each conflict case.

In assessing causes and consequences, this book avoids value judgments about their desirability or the desirability of the conflicts themselves. However, each case explores the implicit and explicit goals of the parties to the conflict and often uses these objectives as a means of evaluating whether the conflict ends up serving the interests of one or more of these participants. In any case, in none of the resource conflicts covered—and, indeed, in no major international resource conflict in recent memory—has there been a clearcut outcome of complete success for any of the parties to the dispute. The outcomes of resource conflict instead appear to be much more mixed. While policy prescriptions for minimizing conflict are an integral part of this study, they are not included in the case analyses and instead receive separate treatment in Chapter 10, because the context for these recommendations is explicitly normative.

SOURCE MATERIALS AND LIMITS TO GENERALIZATION

Because the included cases are relatively recent and highly sensitive, relevant primary sources of information are generally not available. As a result, secondary sources comprise the bulk of the data base. Fortunately, due to the prominence of all of the cases, these secondary sources are quite extensive, usually of high quality, and reasonably insightful regarding the causes and consequences of conflicts over the world's resources. The interpretation of each case always relies on multiple sources of varying ideological slants, which in some cases fundamentally contradict each other. While the accuracy of some of the source materials is not flawless, the reliability of these materials seems unexpectedly high.

Given the research design, sampling scheme, and data base specified, the findings which emerge here have very limited generalizability. Extrapolation of results—whether across time to different historical periods, across space to nations and/or groups not covered, or across issues to concerns other than natural resource conflicts—can be done only with great caution, due to the small number of cases considered and the frailty of the case interpretations. The case analysis is thus oriented toward illustration as much as proof, and the policy implications are presented with considerable tentativeness.

This discussion of methodology paves the way for presentation of the five case studies in the subsequent chapters. While the restrictions on validity, reliability, and generalizability are great, they by no means detract from the fascinating and sometimes startling patterns of resource tensions in the conflicts that follow.

4

Conflict over Endangered Species—the Whaling Confrontation

Of all the resource issues that produce environmental tensions, endangered species involve the most basic emotions. The sentimental and often intensely caring attachment by humans to other life forms (both plants and animals) seems quite understandable, and this attachment often connects with sound ecological concerns about the need to preserve diversity and harmony in nature to maintain global environmental quality of life. Notions such as carrying capacity and maximum sustainable yield play a critical role in gauging environmental deterioration and in designating acceptable levels of exploitation of these renewable resources. But often disagreement may persist about the existing stocks of a species and the rates of regeneration, and this lack of consensus exacerbates tensions between those seeking to protect and those seeking to make use of natural resources.

Whales are, of course, but one of many endangered plants and animals that have attracted widespread attention; tropical rain forests, the rhinoceros, and the manatee are typical of other such concerns by environmentalists. But one distinguishing characteristic of whaling is that whales are a jointly held "common-pool" resource found in the oceans outside of any nation's jurisdiction. Common-pool resources present special problems from a resource conflict perspective: forced sharing of resources can promote international tensions, as mutual agreements are frequently ignored.[1] Ultimately, in an arena of scarcity, pluralism, and paranoia, common-pool resources may turn into hotbeds of competition and resource conflict, far more so than individually held or nationally held resources.[2] The oceans currently exhibit this tendency, primarily over fishing rights, and outer space in the future may display similar proclivities, principally over satellites, planetary mining, or even "Star Wars" weapons systems. In such circumstances, common-pool ownership often intensifies the incentive for short-term overexploitation of resources and their subsequent degradation.[3]

DESCRIPTION OF THE WHALING CONFRONTATION

Although humans have been consuming whaling products since "the earliest days of civilization," the most intense hunting has occurred after World War II.[4] The worldwide whale kill reached its peak of 66,090 in 1962, well above the largest annual catch in the days of traditional Yankee whaling in the nineteenth century.[5] Today's modern fleets are armed with radar, sonar, helicopters, and long-range explosive harpoons, and they travel much farther for their catch and kill younger and smaller whales. Whale products have continued to serve a variety of functions in cosmetics, medicine, clothing, human and pet foods, illumination, and lubrication and cleaning.

Throughout much of whaling history, the ebbs and flows in hunting efforts were due primarily to routinely and incrementally changing circumstances, such as the availability of substitutes for food and energy, the difficulty of finding and catching whales, and the profitability of whaling relative to other enterprises. But in recent decades, there has been a major disruption as interventionist attempts to regulate and curtail whaling began.

International Whaling Commission–Whaling Nation Conflict

The most important governmental body involved in the regulation of global whaling is the International Whaling Commission (IWC), established in 1946 as "the first international body given the power to grant complete protection to endangered species and to set up yearly hunting quotas for the other species."[6] The organization has always contained a mix of whaling and nonwhaling nations, and its membership—which began with fourteen nations—has recently mushroomed to over forty.

Because the IWC's constitution charges it to protect whaling interests as well as whales, commercial whaling concerns have largely dominated the organization since its inception. Due to the lack of reliable data on whale populations (the whaling nations themselves gather most of the information), the IWC has traditionally chosen quotas to suit the business needs of the whalers and has, until recently, set higher quotas for whales killed than the number whaling vessels could actually find. Any restrictions the IWC imposed in its early years "were too little, too late, and were often rendered ineffective by individual vetoes."[7] Member nations may opt out of the commission's decisions simply by lodging an objection within ninety days afterwards, and in any case the IWC has no enforcement power over members. These vetoes usually have come from Japan and the Soviet Union, which in recent years have caught the majority of the world's whales. Instead of fulfilling its two main goals of safeguarding the whales and keeping the whaling industry alive, the IWC "presided for the first 20 years of its existence over the depletion of nearly all the world's whale populations," and the whaling industry "experienced a disorderly, though long drawn-out, collapse."[8]

After the United Nations Conference on the Human Environment passed a resolution in June 1972, calling for a ten-year moratorium on commercial whaling to allow time for more research on whales and the development of a better regulatory system, the IWC (while rejecting the moratorium idea) decided in its annual meeting later that month to allow international observers on whaling ships to ensure compliance with its quotas. Hesitant and faltering steps toward more effective IWC regulation of whaling followed until 1982, when the commission finally voted to phase out all commercial whaling by 1986.

Though both the Soviet Union and Japan promptly filed formal objections, claiming the ban was not based on scientific evidence, both nations have since changed their tune.[9] At the 1985 IWC meeting, the Soviets announced their intention to end commercial whaling, and their last whaling ships finally ceased operations on May 22, 1987.[10] The Soviet Union's compliance resulted from (1) the deteriorating condition of its whaling fleet; and (2) the action by the Reagan administration on April 4, 1985, to cut in half the number of fish the Russians could catch in American territorial waters due to violations by the Soviets of IWC quotas.[11] This penalty imposed by the United States was mandated by Congressional legislation: the Packwood-Magnuson Amendment passed in 1979 required the Secretary of State to take such action against any nation violating directives of organizations like the IWC. Japan also had been violating IWC quotas, and in November 1984 the United States threatened Japan with the same penalty but did not impose it.[12] However, twelve American conservationist groups filed suit to make the U.S. government enforce the penalty on Japan, and they won both the suit and a subsequent appeals ruling.[13] So on April 5, 1985, Japan announced that it would end all commercial whaling by March 1988.[14]

But very recently the whaling nations discovered yet another means of escape from the hunting ban: in the summer of 1986 Iceland, Norway, and South Korea announced plans to continue hunting whales for "scientific research" purposes rather than commercial ends;[15] and in April 1987 Japan stated that it would send a whaling fleet to the Antarctic to kill hundreds of whales for "research purposes" so as to prove that whaling stocks are high enough to lift the global whaling ban.[16] This loophole and others like it (such as exceptions for aboriginal and subsistence whaling) continue to impede the effectiveness of IWC regulations.

Transnational Conservationist Group–Whaling Nation Conflict

Simultaneous with the attempts by the International Whaling Commission, transnational conservationist groups undertook steps using quite different strategies to restrict global whaling. These environmentalist activities began in the 1970s and remained quite separate from regulation efforts of anti-whaling governments. Two principal actions emerged from this private lobbying group effort: (1) a boycott against all products of Japan and the Soviet Union initiated in 1973, and (2) the direct physical interference with whaling ships in international waters initiated in 1975.[17] Both actions are somewhat extreme in the context of international activities by conservationist groups in other resource areas.

The Animal Welfare Institute (AWD) in the United States initiated the boycott of Soviet and Japanese products, and by 1974 the effort had the support of twenty-one conservationist groups with a total of almost five million members, most prominently the National Audubon Society, National Wildlife Federation, Sierra Club, Friends of the Earth, Fund for Animals, and Environmental Defense Fund.[18] Thousands of Americans sent letters and signed statements supporting the boycott,[19] and the Japanese government voiced considerable concern.[20] The boycott centered on Japan because the Soviet Union was less vulnerable, since it sold fewer consumer goods in the United States. However, an evaluation by the Sierra Club's Whale Task Force in the summer of 1976 revealed that the boycott had not reduced Japanese export sales in the United States (which had continued to grow in volume and value) and had actually hurt American-Japanese cooperation on some more critical issues.[21] So in the summer of 1977 most of the major conservationist groups withdrew from the boycott.[22]

The Greenpeace Foundation, a conservationist group founded in Vancouver, British Columbia, in 1970, has used ocean voyages by its small fleet of ships to curtail whaling. These annual voyages cost up to $100,000 a summer, are funded by private contributions, and have focused on the Soviet whaling ships as their primary target because these ships operate much closer to the United States than do the Japanese vessels.[23] Upon approaching a Russian whaling fleet, the conservationists place themselves right between the harpoon guns and the whales in inflatable rubber boats, with the hope of deterring the slaughter. On occasion, the Russians have fired explosive harpoons right over the conservationists to kill a whale, but generally the presence of Greenpeace boats has induced the Russians to abandon their quest at least temporarily and has saved a considerable number of whales in the process.[24] The goals of these Greenpeace voyages are to prevent the killing of whales and focus public attention on the whale slaughter through nonviolent means.

Quite recently, more radical "eco-guerrilla" conservationist groups, most notably the transnational Sea Shepherd Conservation Society, have taken this physical confrontation strategy one step further.[25] Since 1979 Sea Shepherd has used violent means, including ramming and blowing up whaling ships, to protect the whales. Whalers have reacted with anger (and some fear) to this group, calling its members "terrorists" and "kamikaze conservationists," but its overall effectiveness seems difficult to evaluate.

CAUSES OF THE WHALING CONFRONTATION

The origins of the resource conflict between the primary whaling nations (Japan and the Soviet Union) and the anti-whaling organizations (the IWC and the conservationists) lie in vastly different assumptions about the whaling resource. Despite the considerable length of this still ongoing confrontation, mutual antagonisms rather than understandings seem to have increased. The presence of a dwindling common-pool resource supervised by a weak organization (the IWC) characterized by some as a "toothless watchdog" has exacerbated existing tensions.[26]

First, the whaling nations view their own need for the resource as quite high, while those oriented toward restrained whaling view that need as low with many substitution possibilities. Although the Russians have recently used whale meat principally for the purpose of pet food, fertilizer, and livestock feed, they claim that the population in underdeveloped Siberia relies on whales as a cheap source of protein.[27] The Japanese claim that whales play a large role in Japan's cultural heritage and that the Japanese have been eating whale meat since the eighth century.[28] They compare the place of whale in their diet to that of hamburger in the American diet and contend that they would "face starvation" if they gave up whaling.[29] The Japan Whaling Association has bluntly stated that "whales are a resource to be used" and that "in the history of mankind, no country has ever been politically deprived of an entire food source."[30] Furthermore, Japan has asserted that a whaling ban would have devastating effects on its national economy, as whaling is about a fifty-million-dollar-a-year industry directly employing fifteen hundred persons and indirectly responsible for fifty thousand jobs.[31] Conservationists directly challenge these resource views: they stress that abundant and inexpensive substitutes exist for every whale product, and that Japan in reality derives only about one percent of its total protein from whale meat.[32]

Second, whaling nations view their own use of the resource—and consequent responsibility for its scarcity—as low, while advocates of restrained whaling view that use and responsibility as high. Japan and the Soviet Union claim that their whaling industries have recently operated on a quite limited basis: the Soviets since 1975 have operated with just two aging fleets,[33] and Japan's whaling industry (whose six fisheries merged their whaling divisions in 1976) have maintained a whale catch that is just a fraction of what it was in the 1960s.[34] With regard to responsibility, the Soviets did not have a significant share of the world whale catch until after 1959 and the Japanese until after 1946, due in the latter case to American encouragement as a means of providing scarce red meat.[35] So instead these two whaling nations point their fingers at the indiscriminate whaling prior to 1945 by the United States, the United Kingdom, the Netherlands, and Norway.[36] In contrast, conservationists stress that recent whaling activity has been larger and more ominous than in any previous period, with the Japanese and Russians together killing about one hundred whales a day in a one-hundred-million-dollar-a-year business.[37]

Third, the whaling nations view the size of the existing resource stock—and the consequent viability of continued exploitation of the resource—as high, while those seeking to restrain whaling view that stock and viability as low. The differences between these two groups in estimates of whale stocks may vary by as much as five thousand percent.[38] The Japanese and the Soviets regard fears about whales being endangered or becoming extinct as "unscientific," "emotional," and "from people who are unfamiliar with the essence of whaling problems."[39] This skepticism receives support from some scientists who argue that most whale species "are not endangered in the biological sense of the word."[40] Furthermore, Japan's whaling industry claims to have a long-term emphasis on keeping its catch well within the limits for natural reproduction and on "maintaining a balance between preservation and reasonable utilization of whale resources."[41] The Soviet whaling industry claims that whaling can continue because "according to the specialists, the stocks will not dwindle."[42] Conservationists (and scientists supporting their claims) retort that whales are near "the abyss of biological extinction" and that the Japanese and Russian whaling industries have been deliberately "whaling themselves out of business" due to a belief that profits are greater in hunting whales on a large scale for a few years—until several species are extinct—than to do so indefinitely on a limited basis.[43] Environmentalists object to this policy on moral and aesthetic grounds because of whales' size, grace, "singing" ability, and rights as a species, and on scientific grounds because of the research potential from the study of the whale's unique brain and the food potential for two million tons of protein a year from whale meat.[44] Even human survival itself has been brought into question by the scarcity of whales:

> But their [whales']extermination would mean more than just an irretrievable loss to civilization. It could even signal our own destruction. The more than half-million baleen whales that formerly roamed the world's oceans were an inestimable factor regulating the krill-phytoplankton economy of the sea. We now know that the major portion of oxygen in the earth's atmosphere is produced in this complicated ecosystem, and the removal of such a key element as the whale without careful consideration of the consequences could jeopardize the supply of oxygen which supports the life of both man and whale.[45]

Finally, the whaling nations view their own conformity to resource regulations as high, while those seeking to curtail whaling view that conformity as low. The Japanese have stated that they will "abide by, without any reservation, the conclusions reached by the IWC";[46] and the Soviets have asserted that it has observed IWC regulations so faithfully that "Soviet legislation envisages various measures of punishment for violators of the rules, including the institution of criminal proceedings."[47] This stated attitude of compliance is present despite anger by the whaling nations at the motives behind the regulations: Japan has accused the conservationist movement (particularly in the United States) of basing its restrictive pressures on (1) racism toward Japan, (2) a need to vent animosity concerning trade frictions, (2) support from manufacturers who feared Japanese competition and so encouraged a boycott, and (4) a desire "to promote shipments of American cattle and grain to Japan as substitutes for whale meat."[48] The Soviet Union simply characterizes this movement as basing its restrictions on "political considerations."[49] To these claims conservationists have retorted by providing detailed chronicles of actual violations of restrictions by these whaling nations. On many occasions since the IWC began to restrict whaling in 1972, the Japanese and Soviets have disregarded the IWC's quotas, set up their own instead, and killed whales well beneath legal size limits.[50]

CONSEQUENCES OF THE WHALING CONFRONTATION

This resource conflict has had wide-ranging effects. The impact on the resource itself—whales—has generally been to increase the size of the whale stocks through the enhanced restrictions on the whaling industry. Despite continued illegal whaling activities and discoveries of new regulation loopholes, such as whaling for scientific research, there is little doubt that whaling has dramatically decreased since 1972; despite the perilous status of bowhead, humpback, right, and blue whales, several scientists believe that the populations of other whale species are beginning to recover from the devastation of the heaviest period of exploitation.[51]

The impact on the whaling nations has been to reduce their consumption of whale meat sharply and to force economic adjustments in terms of using substitute resources and relocating displaced workers. The Soviet Union appears to have reacted to these changes with a tone of resignation,[52] but the Japanese have responded with an "emotional and political uproar" representing deep-seated anger and indignation, and Japanese whalers are still trying to reverse the decision on the ban, despite the rarity of whale eateries in Japan and the declining public demand for the product there.[53]

As to the advocates of whale conservation, the whaling confrontation has enhanced their prestige and scope of activities. The IWC has grown in membership and in its monitoring activities, and the presence of unilateral legislation, such as in the United States, to penalize violators of IWC agreements has significantly enhanced its credibility. The interference by Greenpeace and to some extent Sea Shepherd has clearly made whaling more costly in terms of time, money, and notoriety, and the dramatic form of the high seas encounters has raised "the environmental consciousness of both the resource consumers and the world at large."[54] Although Greenpeace consulted no governmental authorities to gain permission for its actions and thus lacks a certain kind of international legitimacy, neither the American nor the Canadian government has criticized the group's anti-whaling efforts because it has huge public support, and the Soviet government has remained silent because it does not wish to draw attention to the confrontation.[55] The small scale of Soviet whaling operations and the large support from world public opinion were critical to Greenpeace's success and provided a springboard for expansion of the group's concerns into other environmental issues. Finally, the feared worsening of American-Japanese relations as a direct result of conservationist pressures regarding whaling did not materialize.

5

Conflict over Fossil Fuels—the Oil Crisis

The cornerstone of international tensions over fossil fuels is a pragmatic anxiety about maintaining one's lifestyle, vastly different from the emotionalism of the endangered-species controversies. Although the Third World is facing bleak prospects regarding its fossil fuel needs, the primary arena for concern over this resource issue is the developed nations whose industrial, residential, and transportation systems still depend so heavily on these energy sources. Global consumption of fossil fuels for industrial purposes has doubled practically every decade, while nine-tenths of the people in the poor nations use no fossil fuels at all.[1]

In 1984, of the three fossil fuels, oil produced the largest share of the world's energy (35 percent), coal the next largest (27 percent) and natural gas the smallest (17 percent).[2] Each of the three poses special problems during widespread use: oil has its largest deposits in the volatile Middle East and has caused major ecological damage during accidental spills; coal is dirty and is hazardous and difficult to mine; and natural gas is relatively expensive to produce and transport as a commercial fuel. Being nonrenewable resources, supply shortages in the fossil fuels can be permanently resolved neither through sound resource management, since regeneration is not a possibility, nor through self-reliance strategies, since many nations are simply not geographically blessed with these resources. These limitations appear to make resource conflict more likely and more severe than when dealing with renewables. While global scarcity in the fossil fuels thus seems to be more real than contrived, local/regional scarcities and distributional bottlenecks could still be easily susceptible to political and economic manipulation.

The oil crisis is not the only recent manifestation of international tensions over fossil fuels: for example, in 1981–82, the Soviet natural gas pipeline controversy reached an antagonistic climax, driving a wedge between the United States and its Western European allies over the relative priority of access to fossil fuels and opposition to Communism. But the magnitude of the 1973–74 oil crisis dwarfs not only these other fossil fuel conflicts but indeed all of the other resource conflicts contained in this study. While some people have felt that oil is unique among the fossil fuels in its capacity to generate a severe global resource war, the oil crisis has provided for many a model of the potentialities and limits of the resource weapon in world politics.

DESCRIPTION OF THE OIL CRISIS

Though tensions between the oil-producing and oil-consuming nations had existed for decades, the early 1970s provided special conditions conducive to crisis. The global demand for oil was at an all-time high in the summer of 1973, fueled by a boom in the Western industrialized economies and a switch to oil as a primary energy source. The Arab world had become increasingly dominant as the geographical source of imported oil for Western Europe, Japan, and the United States.[3] Meanwhile, the Organization of Petroleum Exporting Countries (OPEC), founded in 1960 to prevent dips in oil prices, began to grow in bargaining power, and the Arab states—in the subgroup called the Organization of Arab Petroleum Exporting Countries (OAPEC)—began to achieve a stronger unity.[4] This growing strength in OPEC in turn caused Western multinational oil companies, which up to this point had been close to oil-importing nations, to become "much closer partners" with the members of OPEC.[5]

On October 6, 1973, the Yom Kippur War began, as Egypt and Syria simultaneously attacked Israeli-held territory in the Sinai Peninsula and the Golan Heights, and the United States strongly supported Israel. After Saudi Arabia warned the United States that a continued supply of oil would depend on a shift in American policy toward the Arab-Israeli conflict, on October 17 OAPEC met in Kuwait and (with the exception of Iraq) approved an immediate 5 percent cutback in oil production, to be followed by further 5 percent cutbacks until Israel was induced to withdraw its troops from the Arab territories occupied since 1967 and to grant legal rights to the Palestinian people. At the same meeting OPEC as a whole agreed to impose the largest increase in oil prices in its history. Following these decisions, Saudi Arabia and Qatar chose to cut oil production by 10 percent, and Libya and Abu Dhabi embargoed oil exports to the United States. On October 19, President Richard Nixon asked Congress for a $2.2 billion military-aid program to Israel. In response, Saudi Arabia joined in the oil embargo against the United States. Soon all of the other members of OAPEC became part of this embargo, and it was extended to include other nations, most notably the Netherlands, which was the vortex of the European oil distribution system. The West was shocked and angered by the embargo, and the Pentagon even began hinting that the crisis might provoke the American government to use military force in the Middle East.[6]

By December 1973 the price per barrel of oil had quadrupled from its pre-crisis level, but OAPEC had gradually begun to increase oil production and relax its demands for Israeli withdrawal. Finally, due to fear of worldwide financial collapse, fear of American military retaliation, peace moves (signified by Israeli disengagement agreements with Egypt in January 1974 and with Syria in May 1974), internal bickering, and/or frustration about the lack of political effects of the crisis, the Arab states ended the embargo on the United States on March 18, 1974, and on the Netherlands on July 10, 1974, with Saudi Arabia subsequently increasing oil production substantially.[7]

CAUSES OF THE OIL CRISIS

The roots of the resource conflict between OAPEC and the oil-importing nations (primarily the United States, Western Europe, and Japan) lie in the convergence of an economic opportunity for use of a resource weapon with a frustrating political issue—the Arab-Israeli conflict—about which those in control of the resource weapon had strong desires. The appeal of this combination was too great for the conflict initiators to resist, especially because the victims were unable to adapt or respond to the worsening situation.

However, one of the major areas of controversy concerning the origins of the oil crisis relates to the role of the Arab-Israeli war. Many analyze the oil embargo as precipitated by a mixture of political concerns about Israel and economic desires for oil profits, but this conclusion is not universal. Shwadran contends that the Israel issue played virtually no real part in the occurrence of the embargo:

> The embargo was a natural link in the chain of the struggle between oil producers and the concessionaire companies, which after years of stalemate began in February 1971 to progress rapidly in favor of the producers. However, both the Arab producers and the companies—each for their own reasons—advanced the Arab-Israeli conflict in general and the October 1973 war in particular as the real immediate cause for the embargo. This explanation was not only inaccurate but primarily fraudulent.[8]

In contrast, Sampson takes the opposite position and asserts that profit motivation played no conscious role in the initiation of the oil crisis:

> The embargo was conceived solely in the context of the Arab-Israeli conflict: "It had nothing to do with wanting to increase the price of oil," the secretary of OPEC, Ali Atiga, insisted later, "or with increasing the power of the oil producers. It was meant simply to attract the notice of the public in the West to the Israeli question; to get them to ask questions about why we did it."[9]

On balance, it appears that both of these exclusionary views are too extreme and that mixed motives were indeed involved in the embargo. Moreover, on occasion Arab spokesmen would mention still other goals behind the imposition of the embargo, such as to compel other states to provide military and economic aid to the Arabs.[10]

Many forces contributed in the early 1970s to the unprecedented global demand for oil partially responsible for the resource scarcity during the crisis. Vernon points out that since World War II this demand for oil had grown faster than that for any other energy source due to (1) the inexpensiveness of oil (ten or twenty cents a barrel); (2) the soaring costs of coal production in Europe and oil exploration in the United States in the 1950s and 1960s; (3) the "very rare convergence" of industrial growth in Europe, Japan, and the United States in 1972 and 1973; and (4) the restrictions on energy exploitation deriving from concerns about environmental protection, including delays in nuclear power plant operation, efficiency-reducing antipollution controls, and reluctance to disrupt ecological systems by drilling for oil.[11] Unlike previous oil supply downturns during the Suez crisis in 1956 and the Six-Day War in 1967, the oil-consuming nations possessed no substantial standby capacity, existing stockpiles were meager, and few energy substitutes were available for oil.[12] The United States, the Western power most self-reliant in oil, had curtailed its own oil production since 1970, and

> two of the major industrialized regions of the world—Western Europe and Japan—had not only allowed themselves to become almost entirely dependent on OPEC oil for sustaining their economic systems, but had also failed to take any effective counter-action, even in the light of the deteriorating outlook for the viability of their cheap energy policies as a result of the changed post–1970 oil situation.[13]

As Knorr points out, this high inelastic demand combined with high inelastic supply to create the potential for a resource bottleneck.[14] Smart explains that the somewhat surprising unity among OAPEC states in recognizing and applying this potential was because of (1) their geographical concentration in one region—the Middle East and North Africa; (2) the dominant role of a few international companies in this region, providing a common experience and common target for OAPEC; (3) the high contribution of oil production to government revenue and foreign exchange earnings with few labor requirements; (4) the extent to which world oil prices at the time had resisted sudden downward movements and had been rising steadily; and (5) the political consensus within OAPEC regarding the Arab-Israeli conflict.[15] The Arab states not only realized their special position of power but also shared resentment of the unequal and lavish lifestyles of Western consumers and eagerness at the chance to lower such discrepancies. The Arab world claimed "it would do them good" to economize, as "eventually all those children of well-to-do families who have plenty to eat at every meal, who have their own cars, and who act almost as terrorists . . . will have to rethink all these privileges. . . . And they will have to work harder."[16] The major multinational oil companies (the "seven sisters"—Exxon, Gulf, Texaco, Mobil, Socal, British Petroleum, and Shell) "were compelled, at the risk of forfeiting their concessions, to be the instruments of the world-wide cutback in oil" and to be helpless to counteract the OAPEC-created supply shortage.[17] These firms were especially vulnerable to pressures from oil-producing governments, which were more effectively organized than oil-consuming states and were able to provide or withhold long-term supplies of crude oil.

CONSEQUENCES OF THE OIL CRISIS

Considerable disagreement surrounds the extent to which the Arab oil embargo was successful for its initiators. Knorr argues that "the economic power of the OAPEC countries was effective in 1973" in alerting "the industrialized countries to the shift in international economic power that had taken place," "to consequent Arab expectations that this shift should lead to a more cooperative attitude toward Arab goals in its conflict with Israel," and to "the new urgency of the Arab cause and the increase in Arab resources and determination."[18] Yet Licklider contends that "the 1973 Arab use of the oil weapon was a failure," because, in terms of the embargo's stated aims of Israeli withdrawal to its pre–1967 borders and restoration of Palestinian rights, "none of these objectives have been met."[19] While assessment of the outcome clearly depends on the assumed priority of the initiators' goals, it does appear that this resource weapon's primary accomplishments were not the anti-Israeli objectives.

Perhaps the most direct consequences of the oil crisis were economic in nature. The most critical impact was not the shortage of oil supply but rather skyrocketing oil prices.[20] This oil price rise created "hyper-inflation" of prices in general,[21] which in turn helped generate a recession in the Western economies, reducing global production and consumption by about 10 percent.[22] This recession, combined with the high oil prices, precipitated a drop in demand for oil from these economies.[23] Since 1973 oil has not only had its demand fall beneath expectations but also been the source for a diminishing percentage of the world's energy needs.[24] The immediate beneficiaries of these economic shifts were the members of OAPEC, for the huge transfer of money from oil-consuming to oil-producing states provided OAPEC with surpluses of foreign currencies through which it could manipulate international money markets.[25] Pollack at the time noted the specific dangers (not all of which were later realized) in this regard:

> This question of financial instability may turn out to be the biggest of the threats posed by the energy crisis. Recurring upheavals in the foreign exchange markets could trigger protectionism and bring about a severe contraction in world trade, conjuring up visions of another world depression. Accentuating the dangers here are the incredibly large OPEC surpluses that are in prospect and the extraordinary rapidity with which they will accumulate. Existing mechanisms and institutions are simply uninterested in handling international transfers on this order of magnitude.[26]

Aside from the oil-producing states themselves, the oil companies "reaped enormous profits" from the high price of oil, although these benefits were short-lived;[27] Maull notes that higher oil prices "suited" the oil companies, because they then would have enough money to diversify their supply sources by developing new reserves and unconventional sources of oil.[28] Among the oil-consuming nations, the economic impact of the oil crisis was not exactly as OAPEC intended, because (1) some Arab oil "leaked" to the United States despite the embargo, and (2) oil companies diverted Arab oil away from embargoed ports and replaced it with non-Arab oil, thus managing to spread the damage "fairly evenly."[29] Indeed, there is reason to believe that Japan and Western Europe were more affected than the primary target, the United States.[30] In retrospect, because of the evening out of the shortages by oil companies and the relatively quick cessation of the embargo, "it is remarkable how little economic damage was done by the use of the oil weapon."[31] Finally, the oil crisis spurred a desire in these oil-consuming nations to reduce future economic vulnerability to an embargo through "accelerating the development of alternatives to Middle Eastern oil."[32] President Nixon's Project Independence, announced in November 1973 and proclaiming the need for the United States to become self-sufficient in energy by "increasing its drilling and use of other fuels and by massive spending on nuclear research," was perhaps the most extreme of these reactions by oil importers.[33]

Complementing this economic impact was a variety of significant psychological consequences of the oil crisis. A heightened sense of emergency and deep concern over energy issues emerged, but was only temporary, replaced by "complacency" within two years of the oil embargo.[34] Tremendous public antagonism developed within the oil-consuming nations; for example, the American "motorist, accustomed to unlimited fuel for the past three decades, was astonished to find that he was dependent on Arab oil, on 'Six Sheikhs and a Shah'. "[35] However, "it soon became apparent, from opinion polls, that Americans blamed the companies more than the Arabs," especially when the multinational oil firms, whose national loyalty was now very much in question, announced record profits just at the time the oil shortage was most severe.[36] Shwadran reflects this sentiment when he asserts that there "can be no doubt" that the embargo's impact "would have been very limited," had these corporations cooperated with consumer rather than producer countries.[37] In stark contrast to this global resentment, the oil embargo "psychologically gave the Arab oil-producing countries a sense of superiority and importance in both an economic and political sense."[38] The major industrialized nations developed a greater appreciation for their vulnerabilities and limitations, as the superpowers realized their inability to control oil events in the Middle East (a lesson still being learned during American attempts to protect Kuwaiti oil tankers in the Persian Gulf in 1987),[39] and the oil-consuming nations as a whole realized their inability to control their own multinational oil companies.[40] Furthermore, in facing an increasingly clear tradeoff between energy and environmental protection, Sampson points out that, for Western industrialized nations in general and the United States in particular, environmental concerns took a back seat and were largely ignored:

> Far from improving the American environment, the energy

> crisis was now further wrecking it: President Nixon had relaxed the pollution laws, allowing sulfurous fuel to be burned for heating oil; permitted more strip-mining for coal; and allowed the building of the Alaskan pipeline, against the objections of the environmentalists. In the crisis atmosphere, the ecologists carried little weight.[41]

Finally, Smart notices that the oil embargo had a critical psychological effect on the confidence and opportunism of producers of resources other than oil; the crisis added "enormously, by its example, to the inclination of other resource producers to attempt the imitation of OPEC's success."[42]

The political impact of the Arab oil embargo was probably the least transitory. First, regarding political concerns about Israel, Maull contends that "the attitudes of Western Europe, Japan, and the United States toward the Israeli-Arab conflict have changed significantly since 1973," as several states moved from a pro-Israel stance to neutrality or from neutrality to a pro-Arab stance, but that "it seems misleading to attribute these changes solely to the impact of the oil weapon."[43] Licklider more forcefully asserts that OAPEC failed to coerce Israel to change its policies, because the Arab states had no direct influence on Israel and instead had to work through the extremely indirect process not only of linking political ends to economic means but also of using the embargo to influence Europe and Japan to influence the United States to influence Israel to do what the Arabs wanted.[44] Second, the political effects on the relationships among the victims—the oil-importing nations—were mixed. On the positive cooperative side, all of the major oil-consuming nations except France in the Organization for Economic Cooperation and Development (OECD) established the International Energy Agency on November 15, 1974. This "counter-cartel" of buyers has had the goals of resisting OPEC actions and reducing Western dependence on OPEC through the strategies of oil stockpiling, sharing in total production and imports during periods of oil supply cutbacks, conserving energy, and jointly developing alternative energy sources.[45] On the negative divisive side, this and other cooperative attempts among the oil-consuming nations were not particularly effective, because of "irreconcilable differences and cross-purposes among the consuming countries."[46] Maull notes that the splits within the Western alliance and the European Economic Community involved not only differences in interest and in levels of dependence on Arab oil but also mismanagement in dealing with these differences, and he particularly emphasizes the disagreements over energy security between the American emphasis on the superpower balance and the European emphasis on nationalistic aims.[47] Such political divisiveness and infighting was not by any means limited to the oil-importing nations: Shwadran shows that while the oil crisis raised Saudi Arabia up to a leadership position in the Arab world, sharp differences of opinion resulted within OAPEC, as members accused each other of cheating on the embargo agreements and the split widened between economically oriented conservative members and politically oriented radical members.[48]

6

Conflict over Food—the Grain Coercion

In contrast to the fossil fuels necessary to maintain modern industrial lifestyles, food is a staple sustaining life itself. Because it constitutes the most basic of all survival needs (other than perhaps air to breathe or water to drink), food as a tool of international diplomacy "undoubtedly has potentially the most direct and inhumane effects" of all the types of resources.[1] For decades the focus of international food concerns has been the least developed countries.

After centuries of global food shortages, a degree of stability in the supply and price of food occurred in the 1950s and 1960s, thanks principally to the United States and its "bountiful" domestic food policies.[2] But between 1972 and 1974, severe global food shortages and unstable rising prices again reappeared, due in part to an American decision to idle a record amount of cropland to boost farm prices, a Soviet decision to offset a crop shortfall by imports rather than belt-tightening, and a monsoon in the Indian subcontinent.[3] While in the mid–1980s the world again experienced a period of food surplus, regional production trends have begun to diverge sharply: for example, per capita grain production in Africa has fallen by nearly one-seventh since 1969.[4]

Hopkins and Puchala summarize the major problems generated by recent world food conditions as chronic food shortages, undesirable instability in supply and price, insecurity of food imports, low productivity of agriculture, and chronic malnutrition.[5] Food tensions have primarily reflected difficulties in distribution and contrived scarcity (typical of renewable resources) rather than in sufficiency of supply and real scarcity, and this predicament has engendered considerable anger and resentment on the part of starving nations.[6]

Grain is of course but one of the food resources, which include milk and milk products, meat, fish, poultry, fruits, and vegetables; and certainly there have been numerous international food conflicts not specifically concerning grain, such as the "cod wars" between Britain and Iceland in the 1970s and the seemingly endless series of battles between developed and developing countries over the amount and nature of food aid. But Morgan's comment that "grain is the only resource in the world that is even more central to modern civilization than oil" indicates the particular importance of this means of sustenance.[7] Moreover, Barnet points out that even the industrial nations are becoming more dependent on the international grain market.[8] The grain embargoes by the United States, which accounts for about half of the world's grain exports, of the Soviet Union in 1975 and 1980–81 reflect the recognition even by the superpowers of the critical nature of this agricultural commodity.

DESCRIPTION OF THE GRAIN COERCION

The two American grain embargoes under discussion were the culmination of a turbulent history of grain relationships between the superpowers. From 1948 until 1963 the United States had generally refused to sell agricultural commodities to the Soviet bloc, and Peterson explains the basis for this early policy which set the stage for later frictions:

> Begun as an adaptation of hot war controls to a cold war climate, the embargo was justified as an offensive weapon that contributed to the national defense. Government spokesmen declared that a well-nourished communist was a direct threat to the United States, both politically and militarily, and that the United States would arm the enemy by selling commodities to the Soviet Union. Furthermore, they argued, agriculture was the Achilles Heel of communism, and rather than bailing out the Soviets by supple-

> menting their production with American grain, the United States should wait until the Soviet agricultural system—and with it the Soviet state—collapsed in an internal crisis brought on by perpetual shortages. Trade in agricultural products was assumed to be more important to the Soviets than to the United States; hence, trade denial was viewed as a useful Cold War tactic. As the violent anti-communist rhetoric of the late 1940s and early 1950s refined and reinforced these beliefs, trade restrictions became a symbol of American resolve, and this in turn became a significant deterrent in attempts to alter or eliminate the embargo.[9]

By the 1950s the United States experienced a significant agricultural surplus, but strong public antagonism toward the Soviet Union prevented any opening of grain trade until October 1963, following a major Soviet crop failure.

The opening of the Soviet bloc to American grain exports paved the way for what became known as the "Great Grain Robbery" of 1972: the major multinational grain companies (Cargill, Continental, Louis Dreyfus, Bunge, and Andre) sold a billion dollars worth of American grain—more than eleven million tons or over one-third of what the United States normally sold annually to all countries—to the Soviet Union.[10] This sale, which represented the first large-scale Soviet entry into the world grain market,[11] was consistent with political ends pursued by Henry Kissinger in establishing a grain linkage to detente:

> The perennial efforts of the farm bloc and the USDA to sell more grain converged with his desire to engage the Russians on as many fronts as possible, to create linkages between economics and politics, and to give the Soviet Union a long-term vested interest in continuing their 'dialogue' and 'relationship' with the United States.[12]

Because of considerable American resentment of this huge grain transaction, the Soviets—when facing a significant domestic grain shortage in the spring and summer of 1975—attempted to obtain grain more indirectly through European subsidiaries of American grain companies, and this effort met with some limited success. But the United States, buoyed by the successful use of the grain weapon in September 1974 when pledging a hundred thousand tons of wheat to Egypt in exchange for cooperation in Kissinger's Middle East peace plan, felt that Americans could take advantage of Russian vulnerability and apply "the leverage of American 'grain power'" against the Soviet Union.[13] On July 24, 1975, the United States suspended further grain sales to the Soviet Union and sought a Soviet agreement to ship to the United States ten million tons of oil a year at a substantial discount from the OPEC price.[14] The rationale for this American move included President Ford's desire to limit price inflation and his awareness of the Soviet position as the world's largest petroleum producer with a significant surplus of oil.[15] On September 10, the United States extended the embargo to include Poland, which heavily depended on Soviet grain exports and had been buying considerable amounts of grain from American companies during the embargo against the Soviets. But the Russians regarded the embargo as "extortion,"[16] claimed they would "starve to death" before succumbing to such resource blackmail,[17] and were able to wait out considerable domestic hardships, because "the authoritarian nature of the Soviet regime permitted its political leaders to impose such belt-tightening measures in the country with little public outcry."[18] Furthermore, the multinational grain companies continued "scrounging up grain" for the Soviet Union during this period,[19] and angry and well-organized farm protests against the embargo were flaring up in the United States.[20] So on October 20, 1975, the Americans lifted the embargo (the embargo on Poland had ended on October 10) and signed a new long-term trade agreement with the Soviet Union devoid of any of the requested stipulations concerning oil.

The circumstances surrounding the next American grain embargo of the Soviets were slightly different. Once again, because of a disastrous Russian grain harvest in 1979, the Soviet economy appeared vulnerable to disruption, and the Russians had planned to import a record quantity of grain from abroad. The United States had just had a bountiful harvest, and major grain exporters other than the United States were less prepared than usual to meet Soviet import needs.[21] But the immediate trigger of the American action was the Soviet invasion of Afghanistan in December 1979. On January 4, 1980, President Carter, in the "most massive cancellation of grain contracts in history,"[22] suspended American grain sales to the Soviet Union (beyond eight million tons guaranteed in the 1975 bilateral agreement) expressly to "punish" the Soviets for their military occupation of Afghanistan. Paarlberg remarks that "never before had U.S. food exports to the U.S.S.R. been suspended in pursuit of a noncommercial, foreign policy objective."[23] Carter received more domestic support for this sanction than did Ford, because the American government provided extensive compensation to American grain producers affected by the embargo and because the American public felt patriotic in the wake of the Iran hostage crisis as well as the Soviet invasion of Afghanistan.[24] Nevertheless, American farmers soon began to blame the embargo for low grain prices, low incomes, and other difficulties, and the American Farm Bureau withdrew its support for the action in April.[25] As in 1975, Eastern Europe served as a supply leak during the embargo, but Carter could not extend the sanctions this time to include Poland, because Poland had obtained a special bilateral grain agreement with the United States. Although the Soviet Union could have obtained American grain during this period through "transshipment"—moving the resource through third party nations not covered by the embargo—the Russians found it simpler to use premium payments to entice other grain-exporting nations to make temporary increases in grain shipments to the Soviet Union through drawing from surplus stocks and redirecting grain from other potential buyers.[26] The United States tried in vain to restrain other exporters, including Argentina, Australia, and Canada, from falling prey to this Soviet strategy. Even though no support from allies remained for the sanctions and Soviet imports were not being restricted, President Carter continued the sanctions through the rest of his term in office.[27] Finally, in April 1981, President Reagan lifted the grain embargo; afterwards, the United States offered the Soviets up to twenty-three million tons of American grain, but the Russians chose to import only fourteen million tons, as they increasingly relied on other nations for their grain supply.

CAUSES OF THE GRAIN COERCION

Despite the differences in the specific triggers of the 1975 and 1980–81 grain embargoes—a desire in 1975 to obtain large amounts of inexpensive oil and in 1980–81 to inflict punishment for the invasion of Afghanistan—the basic origins were similar of the two coercive applications of American food power on the Soviet Union. While in 1975 the memory of the "Great Grain Robbery" was doubtless much more vivid, both applications occurred in an atmosphere of American resentment of previous Russian success in obtaining large quantities of American grain during times of shortage without paying adequate economic or political costs.

Several underlying motives contributed to the American decision to use grain as a coercive weapon against the Soviet Union. The oil crisis of 1973–74 caused American foreign policy makers (particularly Henry Kissinger) to feel "that grain was a national resource that was simply too important to be left to the devices of the marketplace or to the dealings of grain traders."[28] Proponents of "food power," armed with a supportive 1974 Central Intelligence Agency study, have suggested in the 1970s that grain "shortages may supply the United States with a potent weapon in world politics, allowing America to regain the influential position in world affairs that has eroded in the post-Vietnam era."[29] In Butler's view, American grain coercion "owed more to the desperation of the search for some device to reassert 'primacy in world affairs' than to cool, logical evaluation."[30]

Wallensteen more generally identifies the three fundamental objectives of American foreign grain policies as economic interest in disposing of agricultural surpluses, strategic interest in combatting Communism, and humanitarian interest in reducing poverty; he contends that the first two goals usually are dominant, as was the case in the grain embargoes of the Soviet Union.[31] But when American foreign grain policies deal specifically with the Soviets, face-saving becomes particularly prominent. In the second grain embargo, for example, President Carter felt strong pressure to demonstrate at home and abroad that he could stand up to the Russians, making the coercion "an unhappy necessity."[32]

Despite great internal criticism of the American grain embargoes from farmers and others hurt economically by these sanctions, there seemed to be an intrinsic popular appeal in the United States using its most potent resource weapon to achieve foreign policy ends. After the oil crisis, many Americans were swept along by "we freeze, they starve" and "no crude, no food" slogans and only detailed analysis in the Treasury and State Departments of the ramifications of a grain embargo against the Arabs prevented a move in this direction.[33] Furthermore, the Soviet Union appeared to be the most feasible and desirable target for such an embargo, combining heavy dependence on imported grain with ideological opposition.[34]

Finally, this resource conflict was the product of an absence of control on two levels: at the international system level over the actions of individual states and at the national government level over the actions of private multinational corporations. The first issue reflected the failure of the World Food Conference, held in Rome in November 1974, to establish meaningful provisions dealing with international food security; this anarchistic situation was highly conducive to unilateral uses of the grain weapon for political ends.[35] The second issue reflected a growing recognition by the American government that "it had little power to control what Cargill, Continental, or Cook did in Europe, Australia, or Argentina."[36] The close relationship between the multinational grain companies and grain-importing nations (which is in contrast to the closeness of oil companies to oil-exporting nations) was critical in causing the conflict that ensued after the imposition of the American embargoes, because this relationship enhanced Soviet confidence and flexibility and thus increased Russian resolve to resist American challenges.

CONSEQUENCES OF THE GRAIN COERCION

It is generally conceded that both American grain embargoes of the Soviet Union ended in failure. Wallensteen contends that, for food to be used successfully as a political weapon, the necessary conditions are (1) scarcity of the food resource in terms of the supply/demand ratio; (2) supply concentration in the hands of a few producers or sellers; (3) demand dispersion in which many buyers compete; and (4) action independence on the part of the nation applying the food weapon.[37] At the time of the two embargoes in 1975 and 1980–81, there was neither a severe global shortage of grain nor a supply concentrated in a few hands. In a parallel manner, Paarlberg asserts that an effective exercise of food power requires that (1) within the political system of the initiating nation, foreign policy officials must be able to control the volume and direction of food exports; (2) within the international food trading system, other countries and multinational corporations must be prevented from leaking embargoed grain into the target nation; and (3) within the political and economic system of the target nation, the intended reduction in food imports must be adequate to produce desired impacts.[38] The 1975 embargo failed to meet all three criteria, and the 1980–81 embargo did likewise, except for initial success regarding the first point of control within the American political system.

As Hopkins and Puchala note, the prevailing international system does not appear vulnerable to food pressures: most nations are almost totally self-sufficient in food, farmers cannot easily forego the export earnings of food, and the global availability of alternative supplies from alternative sources in the open food-trading system makes an effective food embargo "practically impossible, as well as morally repugnant."[39] Regarding this last point, Butler argues that the inescapable moral doubts involved in denying international grain supplies, which hits the poorest the hardest, dooms such policies from the outset, particularly for a nation that has lofty foreign policy ideals, such as the United States.[40] From these general considerations, Christensen concludes that "selective embargoes, like that imposed on grain exports to the USSR in 1975, are difficult to fine tune.... [and] will involve public controversy, economic disruption, and conflict with other food goals."[41]

More specifically, the United States, rather than the Soviet Union, suffered painful domestic consequences from the grain embargoes; as Gilmore wrote at the time, "It will take years to recover from the economic and political consequences of the 1980 embargo against the Soviet Union."[42] Estimates of the costs to American taxpayers of the 1980–81 embargo were between two and three billion dollars,[43] although American grain producers exaggerated the damage the embargo generated to their export opportunities.[44] In the 1975 embargo, farm groups similarly claimed two billion dollars worth of damage in terms of lower food prices at home and lost business abroad.[45] Farmers were so enraged about this price decline that some sued the federal government for damages resulting from the embargo.[46] As indicated earlier, the Soviet Union escaped the intended impact of the first grain embargo through transshipment of grain by the multinational corporations by way of third party states, and that of the second grain embargo through direct grain imports from other national suppliers. So the Soviets did not experience significant economic hardships as a result of the first embargo, and the second embargo had even more "minimal" economic consequences due to an excellent 1980 Russian grain harvest.[47] Moreover, the Soviets developed a variety of methods—including finding alternative supplies, increasing domestic production, and reducing domestic consumption—to deal with grain shortages.[48] There is no evidence that the second embargo "had any impact on Soviet policy toward its occupation of Afghanistan." Instead, the sanctions may have caused the Russians to press harder to achieve agricultural self-sufficiency.[49]

Beyond these direct and immediate impacts of the embargoes on the United States and the Soviet Union, there were more indirect, far-reaching consequences. First, the defeat of both Ford and Carter in the subsequent American presidential elections has been at least partially attributed to the grain embargoes.[50] Second, the Soviets have been able to capitalize on the American grain coercion in the arena of international public opinion, labeling American attempts to deprive the Russian people of food as "shameful"[51] and helping to "damage America's reputation as a source of secure supplies."[52] Third, the cumulative impact of the two American grain embargoes has been the Soviet view of the United States as an unreliable supplier with grain exports "tied to foreign policy," this reducing Russian dependence on American grain.[53] Only quite recently have the Soviets again returned to the American grain market (despite high price levels) in "an effort to keep commercial ties open," recognizing "that a stable, long-term relationship with the world's largest exporter is essential in view of the U.S.S.R.'s enormous requirements."[54]

7

Conflict over Nonfuel Minerals—the Strategic Minerals Threat

Of all the types of resources covered in this book, nonfuel minerals are the least publicized and the least widely understood. The absence of awareness exists because the vast majority of consumers do not use most of these substances in their pure form and are not knowledgeable about their contribution to industry and military defense.[1] Nonetheless, the industrialized Western nations are far more heavily dependent on foreign sources for a number of nonfuel minerals than for crude oil, and in the 1980s these minerals have been attracting more attention than ever before.[2] As Varon and Takeuchi note, "Non-fuel minerals can be deeply significant for individual countries, for the overall balance of economic power in the world, [and] for the welfare of very large numbers of people."[3] As with the fossil fuels, these nonrenewable nonfuel minerals are much more the focus of developed rather than developing country concerns.

Of the scores of varieties of nonfuel minerals, the most internationally important ones are iron ore, bauxite, copper, manganese ore, lead, nickel, phosphate rock, zinc, and tin, which together account for an overwhelming proportion of the estimated value of world production of all nonfuel minerals.[4] The global supply of these most prominent industrial materials "seems almost inexhaustible," with the primary problem (like food) being one of distribution as industrialized nations become increasingly dependent upon imported nonfuel minerals from the Third World.[5] The extremely uneven international distribution of these minerals, in which reserves and production capacity for several are concentrated in a few countries, is highly conducive to contrived rather than real scarcity. But since 1950 more of these minerals have been mined than in all previous history.

The mechanization of war has led to a special concern for a particular subcategory of nonfuel minerals commonly called "strategic" minerals, usually defined as those minerals that are deemed essential for defense and that are ordinarily acquired from foreign sources, and thus subject to supply disruptions during wartime.[6] From a national military standpoint, these minerals would be the focus of attention regarding modern weapons needs and the primary source of mineral-related resource tensions.

Since 1980, this defense-affiliated concern about strategic minerals has become particularly acute in the United States, which, though extremely well-endowed in nonfuel minerals, has experienced growing dependence on imports in recent years. This American concern has primarily revolved around four strategic minerals: chromium, cobalt, manganese, and platinum group metals. The United States relies on imports for over ninety percent of its supply of these minerals (compared to only about thirty percent for oil), and the majority of these imports come from South Africa, Zaire, Zambia, and Zimbabwe.[7] Indeed, in this context South Africa has been called the "Saudi Arabia of non-oil minerals."[8] Western Europe and Japan appear to be even more vulnerable than the United States to potential disruptions in strategic minerals.[9]

The resource conflict over strategic minerals differs from all others in this study in that it revolves around a perceived threat rather than a concrete resource disruption or application of a resource weapon. While in the general area of nonfuel minerals, frequent tangible battles have erupted over, for example, control of minerals on the ocean floor or those located in and around national boundaries, the low-level conflict between the United States and the Soviet Union since 1980 over the threat of disruption to American minerals supply has affected governmental policy at least as much as these other cases.

DESCRIPTION OF THE STRATEGIC MINERALS THREAT

Although strategic minerals have been high in prior historical periods, such as between the two world wars,[10] in the 1980s strategic minerals have achieved unprecedented prominence and have become key considerations in diplomacy and foreign policy.[11] This attention has been largely nationalistic, oriented toward energy security, and centered on potential embargoes and supply cutoffs from one or more mineral-exporting nations operating noncollusively, as early anxieties about the development of strategic minerals cartels have faded.[12]

Southern Africa has seemed recently to be a particularly inopportune part of the world for American dependence on strategic minerals. As an Undersecretary of State recently remarked at a conference on strategic minerals, "There is an uncomfortably high degree of overlap between areas outside the United States which are principal resource producers and areas of substantial political instability."[13] Blechman notes that stability in the areas of Southern and Central Africa exporting strategic minerals deteriorated substantially in the late 1970s and early 1980s, as military conflicts have grown within many of these nations and political relationships have become badly "frayed"; consequently, there seems to be a substantial risk of further mineral disruptions—amounting to perhaps 20 to 50 percent of a single year's exports from the region over the span of the 1980s.[14] Shafer asserts that "the most probable threat to the supply of strategic minerals" for the United States and the West is "unplanned, unpredictable, chaotic interruptions of production or supply" through political instability in these mineral-exporting states.[15] Weston points out that ambiguities inherent in political risk assessment make forecasting these supply disruptions difficult.[16]

Recent turmoil in Zaire, from which about ninety percent of the world's cobalt is extracted, provides a tangible example of the kind of instability-induced conflict over resources expected in the future.[17] Beginning in 1964, General Mobutu's corrupt and repressive political regime controlled Zaire, and he received support from the United States and NATO members mainly because of his cooperation with Western mineral interests. In 1977, the FNLC (National Front of the Liberation of the Congo) attacked the mineral-rich Shaba province and demanded that Mobutu be replaced. In response, France airlifted Moroccan troops to thwart the invasion. In 1978 the rebels again attacked, flooding some mines and driving away white technicians, and French, Belgian, and Moroccan troops flew in again to save Mobutu. This last attack created a limited disruption in the production of cobalt in Zaire in 1978–79, with severe temporary effects on importing states.[18] While these incidents are certainly among the most extreme, the potential for instability-related disruption seems high throughout the region, as the ongoing turmoil in South Africa vividly demonstrates.

The timing of this increased American dependence on mineral imports also appears to be inopportune. As Haglund explains, the rapid decolonization movement in the decades following World War II and the declining hegemony of the United States have combined to foster a growing assertiveness among the governments of developing countries "attempting to extract more economic rent from multinational minerals producers operating within their boundaries."[19] As a consequence, Western observers anticipate that disruptions in the supply of strategic minerals may result from not only unintended political instability but also intentional restrictions by Third World governments anxious "to assert their 'resource power' " and "to exert greater control of their mineral wealth."[20]

Compounding and in some minds eclipsing the threat of supply disruptions emanating from the mineral-exporting states is the threat of Soviet intervention. The precise nature of the Soviet threat to Western strategic minerals supply is the subject of heated controversy. The Soviet Union has had much greater self-sufficiency in strategic minerals than the Western industrialized states and has traditionally been a major exporter of strategic minerals; for example, it is the third-largest producer of chromium in the world, and recently even the United States imported over 30 percent of its needs in this mineral from the Soviets.[21] The greatest fears emanate from those perceiving an intentional Soviet strategy of "resource denial" with respect to strategic minerals. A prime advocate of this position is former United States Representative James Santini, chair of the House Subcommittee on Mines and Mining from 1974 to 1982 and later a member of the National Strategic Materials and Minerals Program Advisory Committee, who believes that "a resource war" is in progress between the United States and the Soviet Union:

> The U.S. is dependent on southern Africa for approximately three-fourths of its supply of manganese, cobalt, and chrome. If the Soviet Union knocks off Zaire, if Zimbabwe goes down the drain—Zambia is already a Marxist state—and if South Africa has a racial revolution or invasion or

> both, Russia will control 80% of the market place of these three strategic minerals.[22]

Former President Nixon has quoted former Soviet Premier Leonid Brezhnev as stating, "It is our intention to deprive the West of its two main treasure troves: the oil fields of the Persian Gulf and the strategic mineral resources of central and southern Africa," making Western Europe a hostage to the Soviet Union and isolating the United States.[23] The willingness of the Soviets to impose a crippling embargo on strategic minerals against China, as a result of the Sino-Soviet rift in the early 1960s, serves for some as a stark reminder of the probabilities of such action against the West.[24] Undersecretary of State Schneider explains that while the Soviet strategy "directly associated with the denial of resources to the West and Japan" is not the only interest the Soviets have in mineral-exporting regions, it is an "important motivation" for them to disrupt political life in Southern Africa by using violence and guerrilla conflict to deter Western mineral investment there.[25] The underlying assumption behind these views is that the Soviets have been attempting to gain control over strategic minerals in Southern Africa specifically in order to deny the West access to them.

Other analysts, who have had less policy influence in the 1980s, severely question these alarmist views on the Soviet strategic minerals threat. Shafer summarizes his skepticism:

> Although Moscow no doubt sees benefits in controlling these resources, the prospects for such a grab must seem dim to Soviet planners. Zaire, Zambia, and Botswana are politically fragile, but currently they remain firmly in the Western camp without significant pro-Soviet opposition groups. Moreover, even if the southern African minerals producers passed into the Soviet camp, they would remain dependent on the West as the only available hard currency market for their products.[26]

Ogunbadejo adds that the resource denial theory is weak because the African states (especially South Africa) desperately need to export strategic minerals, Western economic retaliation could occur, and the West could obtain the minerals through transshipment through third parties if denied direct access.[27]

In a more moderate position, Legvold agrees that "simply no evidence exists suggesting that Soviet leaders think in terms of strangling the West by denying it strategic non-fuel minerals in peacetime," but he does argue that the Soviets would use the resource denial strategy during wartime.[28] Furthermore, he asserts that the Soviet Union is encouraging the mineral-exporting states to use their resource leverage "to master their economic destiny" and reduce the power the United States and other capitalist nations have within the global economy: "In a loose and uncertain way, the Soviets may well believe that aggressive economic policies on the part of mineral-exporting countries contribute to the long-term transformation of a Western-dominated international economic order." This last view appears to be the most sensible, especially with reference to the Gorbachev regime but, regardless of the reality of Soviet intentions concerning strategic minerals in Southern Africa, the perception of threat alone by the United States has seemed sufficient to generate conflict.

The prospects of a major resource war resulting from these strategic minerals threats appear remote, but proxy clashes between superpower puppets seem to be real possibilities.[29] An economist at the Army War College perhaps best summarizes the upcoming scenario:

> As we progress into the 1980s and beyond, increasing world demand for high-grade mineral deposits, and their corresponding depletion, can be expected to intensify competition for the world's strategic minerals. A growing militance among the less-developed supplier nations in their quest for a new world economic order, and the simultaneous exercise of Soviet power and influence in the continuing East-West struggle, will further disturb the increasingly competitive environment and heighten its potential for generating international conflict.[30]

CAUSES OF THE STRATEGIC MINERALS THREAT

The fundamental source of the resource conflict between the United States and the Soviet Union over strategic minerals is geopolitics, in the sense that a struggle exists over security of access to geographically concentrated raw materials critical to international power and influence.[31] Haglund points out that "it is only in the 20th century that minerals have appeared as a reason for, not merely a means of, fighting," and that the experiences of the two world wars "underscored the strategic relevance of minerals in modern wars of attrition."[32] Aside from the previously mentioned uneven, unstable, and politically-manipulated supply patterns, the peculiar origins of American and Soviet trends in mineral demand help to account for the bottlenecks in strategic minerals access.

Van Rensburg explains that the growing American dependence on strategic minerals is due to increases in consumption, an expanded range of minerals required by modern industries, depletion of high-grade domestic deposits, neglect of mining and metallurgical research, growing concern with environmental protection and more stringent environmental regulations since 1969, restrictions on exploration and mining in the federal domain, inadequate encouragement from domestic tax laws for minerals exploration, sharp increases in transportation and processing costs since the oil price shocks of 1974 and 1980, and growing competition for available supplies from other non-Communist industrial nations.[33] Ogunbadejo exemplifies the impact of modernization on strategic minerals dependence by noting that the space shuttle alone uses more than twenty of these minerals, and he further contends that the Reagan administration has placed a special emphasis on this minerals dependence to rationalize the American government's staunch pro–South-Africa and militant anti-Soviet stance.[34]

Many analysts assert that this American dependence on external supply of strategic minerals is particularly problematic because of the limited possibilities for substitution and recycling: the development time for substitution systems may be ten years for complex weapons systems, and recycling possibilities appear insignificant as a percentage of demand for most strategic minerals.[35] The United States contains few known reserves of many strategic minerals.[36] Furthermore, enhancing supply flexibility through deep sea mining—especially of manganese nodules of the ocean floor containing manganese, nickel, copper, and cobalt—is no longer an option for the United States: the Reagan administration withdrew the United States from the Law of the Sea Treaty, largely because of an inadequate legal basis in the treaty for the encouragement of American mining companies to invest in the high costs of sophisticated exploration, ships, and processing facilities offshore.[37]

However, Shafer challenges the "myth" of American mineral dependence, claiming that "minerals cost much less and are used in far smaller quantities per unit of final output than oil and are easier to store"; that global reserves of strategic minerals are not fixed but instead expand depending on legal provisions, mining and processing technology, price, and geological knowledge; and that "the United States could lose a substantial portion of its strategic minerals imports without facing any threat to its national security" because many of these materials contribute to nonessential products or could be replaced with available substitutes.[38] However, he does admit that the vulnerability of the European Economic Community and Japan poses a great strategic minerals threat for the United States, and thus primarily advocates an altered focus of concern for strategic minerals threats, rather than a reduction of the concern itself.

Beyond this precipitant of increasing American minerals dependence, preliminary but ominous indications have emerged that Soviet self-sufficiency in strategic minerals has declined. Recent reports have pointed to increasing Soviet imports of chromium, manganese, copper, cobalt, lead, and bauxite and have concluded that the Soviet Union was approaching the depletion of several of her energy and mineral resources.[39] These reports further argue that, like the West, the Soviet Union would soon become dependent on imports of these commodities and, lacking sufficient hard currency to get them through trade, might well resort to force to obtain them. If these controversial reports hold true, Soviet initiation of this resource conflict could be motivated by economic as well as ideological issues.

CONSEQUENCES OF THE STRATEGIC MINERALS THREAT

Because this resource conflict is characterized more by perceived threats than concrete antagonistic actions, its consequences are somewhat conjectural in nature. But there is rough consensus that a conflict over strategic minerals involving a significant supply disruption for even a short period of time would generate crisis conditions in much of the industrialized world.[40]

The impact of such a conflict on mineral-importing nations would be severe "economic dislocation," with the degree of dislocation depending on the size of the supply reduction, the duration of the interruption, and the extent to which public and private stocks of the minerals in question can offset the shortfall.[41] Furthermore, such minerals conflicts could cause within-state battles reflecting "an increasing tension between resource development and environmental and conservationist ideas," as the pressures to intensify domestic mining escalate.[42] Turning to mineral-exporting nations, the primary effects of this resource conflict would be mixed, as Ogunbadejo suggests: the international power of mineral-rich African nations like Zaire might increase, and strategic minerals could generate needed income for local economies. On the other hand, the increased competition for these minerals could cause further exploitation of African labor and propping up of despots by the great powers and their multinational corporations, and it could further widen the gap in these developing states between an expanding modern mining sector tied to the needs of advanced capitalist economies and a stagnant agricultural sector.[43] Attempts by minerals-importing nations to place embargoes on the export of oil and other products to mineral-exporting nations, as has been considered regarding South Africa, would likely generate counter-embargoes on the export of strategic minerals from affected minerals-exporting states.[44]

This resource conflict would also have consequences for transnational groups. Van Rensburg contends that the propensity of exporters of strategic minerals to consult with one another, to support the New International Economic Order, and/or to adopt overly nationalistic or socialist policies could deter investment by multinational mining companies.[45] Blechman asserts that terrorists or disaffected subnational groups could well take advantage of the opportunity to disrupt an increasingly fragile strategic-minerals production system.[46] Transnational protests and even militant intervention by anti-apartheid groups opposing the mineral-related support by the U.S. government of the South African regime are also possibilities here.

Finally, a concern exists about the impact of such a conflict specifically on American resource security. McMichael paints a gloomy picture:

> Without the minerals it needs, the United States cannot maintain its industrial infrastructure and thereby its economic strength; and without the latter the nation cannot hope to retain its geopolitical strength and independence which (to close the circle) it must have in order to ensure its access to foreign supplies of minerals. Thus, the United States must deal with a circle of vulnerabilities that link together virtually every aspect of its domestic and international life.[47]

In response to the increasing volatility of the global economy and fear of supply disruptions, the Reagan administration established the National Strategic Materials and Minerals Program Advisory Committee to help manage the problem. In addition, the United States has begun to try to reduce its external minerals dependence by establishing strategic stockpiles of a range of minerals not readily available from domestic sources, but this move has angered some mineral-exporting nations.[48]

8

Conflict over Pollution—the Chernobyl Nuclear Disaster

Pollution, the most repugnant of all resource issues, poses unique problems from a conflict perspective. Pollution can range from being an ugly nuisance to a health hazard to a cause of death, and the more serious forms seem to have the closest link to resource conflict.[1] Through pollution, neglect, and overexploitation, "over the past century, human beings have perhaps done more irreparable harm to the natural environment essential to their survival than the combined impact of [all] earlier generations."[2] Greater resource scarcity may generate greater pollution due to (1) the unavailability of resources for pollution control (or the infeasibility of increasing the already high prices of products by adding pollution-control devices); and (2) the need to rely on the most efficient production processes regardless of long-term pollution effects.[3] But this amplification of pollution seems to occur only up to a point; after the onset of extreme scarcity, the rapid decline in production and consumption may reduce pollution levels.

The kind of pollution problems experienced by the rich and poor nations differ: developed nations have a larger problem with inorganic industrial waste from nonrenewable resources, while developing nations have a greater problem with organic human and animal waste from renewable resources. This distinction makes developed nations feel more concerned about and threatened by pollution, since their primary form of pollution is not biodegradable and can have a massive and unpredictable effect on human life.

Assessing the impact of pollution on resource conflict is facilitated by findings of physical discomfort theories, which indicate that unfavorable and oppressive environmental conditions create irritability and produce aggression;[4] and by the widespread recognition that the increasingly frequent pollution spillovers across national boundaries facilitate conflict, with emotional accusations and haggling over accountability.[5] However, extremely high levels of pollution tend to inhibit purposive (goal-oriented and premeditated) conflict, because they seem to make any form of organized human activity difficult. Because pollution engenders difficulties in common-pool resource management due to massive spillover effects, assigning blame or harboring hostility for pollution becomes a futile exercise after a certain level of planetary deterioration. Increased pollution may be a consequence as well as a cause of conflict, as war often pollutes air and water and degrades the soil.[6]

Pollution differs from the other resource issues considered in this book in that its impact on resource availability is indirect: there is usually no problem in the immediate supply of the resource generating the pollution—indeed, an overabundance may exist—and instead there is an inadequate supply of unpolluted resources in an uncontaminated natural environment in areas hit by pollution. Given the tangible (though often invisible) ecological effects of pollution and its uncontrollability, the resulting scarcity tends to be real rather than contrived.

Contamination from the use of nuclear energy is but one of many tension-creating manifestations of pollution evident internationally in recent decades: others include pollution from commercial poisonous chemical leaks, such as Union Carbide's accident in Bhopal, India, which killed more than two thousand people and injured more than one hundred thousand others; major oil spills on the high seas, mainly from tankers transporting the resource; and acid rain, which has changed from being a local urban problem to a regional and global one. But no pollution issue has spawned a greater public outcry than nuclear radiation, and the safety issues surrounding nuclear energy have "aroused spontaneous citizens' movements in every industrial country outside of the socialist world."[7] Heated debates persist about the hazards, waste disposal, and even costs of nuclear energy, and nuclear energy installations across the globe have proven not to be immune to bombings, acts of sabotage, or even direct military attacks.[8] Meanwhile, the overall scale of ongoing nuclear energy production is sizable—as of mid–1986, the world had 366 nuclear power plants in operation for a generation capacity of 255,670 megawatts, representing about 15 percent of global electricity production and an investment of more than $200 billion.[9] The Chernobyl disaster of 1986 brought all of the tensions to a head surrounding this resource.

DESCRIPTION OF THE CHERNOBYL NUCLEAR DISASTER

On April 26, 1986, two powerful explosions destroyed one of the four nuclear power reactors in the Chernobyl energy complex in the Soviet Union, which is the third largest producer of nuclear energy and relies on this energy for over ten percent of its power.[10] The reactor involved in the accident had begun operation in 1983, and it was based on a graphite-moderated design considerably more volatile than the light-water-moderated designs common to nuclear reactors in the West.

The disaster resulted from gross operator incompetence compounded by the poor reactor design.[11] The immediate trigger was a desire by the reactor operators to test safety systems to determine whether the power plant's turbine generators could provide electricity for vital functions during the period between an internal power failure and activation of emergency diesel generators. Previous tests at Chernobyl in 1982 and 1984 had failed, but the specific purpose of the 1986 test was to examine the effectiveness of a new device designed to keep the voltage up during this critical period. This test thus resembled an accident simulation, and the operators scheduled it for April 25, because that was the date when the reactor was scheduled to be brought down for routine maintenance. The chief engineer was on vacation, and his deputy was the principal proponent of breaking operational rules to perform the unauthorized experiment during shutdown.[12] At 1:00 P.M. operators reduced power generation by 50 percent and shut down one of the two turbines, and at 2:00 operators began to deactivate emergency core-cooling pumps to prevent automatic safety features from interfering with the experiment. After a lengthy interruption for an unexpected demand for electricity, the reactor became increasingly difficult to control as power declined too rapidly. The plant operators failed to follow standard low-power operating procedures, ignored dire computer warnings, and continued to shut down more safety systems so that the experiment could continue. At 1:23 AM the operators cut off the steam supply to the remaining turbine generator; power levels and reactor instability escalated, and two massive steam explosions ensued, ripping off the roof of the reactor and ejecting a quarter of the graphite in the core. Numerous fires ignited in the nuclear power complex, and only speedy action by firefighters prevented other reactors from exploding as well.

More than two weeks elapsed before the inferno in the burning reactor was brought under control, entombed under 300,000 tons of concrete and six thousand tons of metal, and in the end the plant released between 3 and 4 percent of its total core inventory of radioactive materials.[13] The direct costs of the accident were thirty-one deaths, one thousand immediate injuries, five hundred people hospitalized, 135,000 people evacuated from their homes in the Ukraine, and at least three billion dollars of direct financial losses.[14] Health-threatening levels of radioactive materials, whose duration in the atmosphere ranged from a few hours to 24,000 years, spread more than two thousand kilometers from the plant in at least twenty nations.[15] Hardest hit (outside of the Soviet Union) was Europe, whose soil, water and crops received as much as five hundred times their normal level of radioactivity.[16] Pripyat, a Russian town close to Chernobyl containing 49,000 people, will be uninhabitable through at least 1991, and between two hundred and 100,000 people will die of cancer in the Soviet Union and Europe due to contamination from the accident.[17] As Flavin concludes, "The Chernobyl accident was by any measure the most serious nuclear accident the world has ever suffered."[18] Once Europe recovered from its shock about the human and ecological damage following the disaster, conflicts began to develop centering around nuclear power between affected nations and the Soviet Union and among neighboring nations with differing nuclear policies.

CAUSES OF THE CHERNOBYL DISASTER

While the preceding description establishes faulty reactor design and operator error as the causes of the nuclear power plant accident at Chernobyl, there were separate reasons why this disruption generated resource conflict. As with most instances of pollution spillovers, the origins of the disaster and the ensuing conflict were neither premeditated nor intentional. Strife occurred simultaneously on many fronts: (1) between the Soviet Union and West European nations, (2) within the Soviet Union, (3) among West European nations, and (4) within the states of Western Europe.

The root cause of the conflict between the Soviet Union and Western Europe was resentment of Soviet information policy and the damage generated by the accident. Flavin explains the Western European frustration with Soviet announcements about Chernobyl:

> Despite an avowed policy of openness, Soviet authorities waited almost three days before announcing the disaster, and that was only in response to outcries from Scandinavia. Even then, authorities downplayed the seriousness, refusing to release detailed fallout information and squeezing stories about the accident onto the back pages of *Pravda*. Although the Chernobyl reactor was blown to pieces in the initial seconds of the accident, days later the Soviet Union was assuring the world that the reactor was "under control."[19]

In contrast to some earlier Soviet attempts to avoid releasing information about incidents with international implications, the contamination from the Chernobyl accident was "immediately detected by neighboring countries by artificial satellite intelligence;" and "West European governments demanded, and East European governments requested, information about the potential danger to their populations."[20] Some Western governments became so upset that they planned to present damage claims against the Soviet Union at the International Court of Justice, but the Soviets assert that they will not accept the court's jurisdiction, for they "flatly deny that Western countries have suffered significant damage."[21] The Soviet Union is angry itself with Western Europe over Chernobyl because, in the disaster's early stages, Western radio stations transmitted speculative and distorted information about the accident to the Soviet public; one top Soviet official has even suggested that Western European governments should compensate the Russians for the damage that these exaggerated press reports did in the Soviet Union.[22]

The roots of the conflict within the Soviet Union were similar—

resentment of the scope of the damage and inadequate information—but were compounded by unhappiness with the sluggishness of the Soviet government in taking corrective action. Only one month before the Chernobyl disaster, a "well-informed and harsh indictment" of conditions at the plant had appeared in a Ukrainian newspaper and been apparently ignored by Russian authorities:

> Local resident Lyubov Kovalevska described in detail "a huge number of unsolved problems," workers' "indignation and, later, desperation," "the low quality of the design and costing documentation," "weakened discipline and responsibility," "acute and apparent shortcomings in the building process," "lack of organization," "defective material," "disruptions in supply," "constantly violated contract obligations," "shoddy structures," "spoilage," "undelivered" goods, and so on.[23]

Hoffmann argues that the Soviet people had "good reason to be apprehensive about the short- and long-term health consequences of the Chernobyl accident," a fear enhanced by "the paucity and timing of the information they have received and the conflicting statements by officials and journalists"; for example, a Russian father of three sent a letter to Gorbachev and the Western press protesting local authorities' "criminal negligence" in not evacuating his family and the town of Pripyat until at least thirty-six hours after the accident.[24] Moreover, the frustrations of the Russian people could only be amplified by power brownouts and reduced industrial capacity and electricity generation fostered by the Chernobyl accident.[25] A particularly ripe arena for internal conflict appears to be between ethnic Ukrainians and the predominantly Great Russian government in Moscow.[26]

The special locational pattern of nuclear power plants has helped to explain why the Chernobyl explosion escalated conflict among Western European nations. Flavin elaborates on this source of conflict:

> Nuclear power has now emerged as an important bilateral issue causing tensions between neighboring countries. Chernobyl demonstrated that the effects of a nuclear accident can cross international borders with impugnity, and in Eu-

> rope, 119 nuclear power plants are located within 100 kilometers (62 miles) of a national frontier. Nuclear plants are often clustered near borders in part because the large rivers that commonly form national boundaries can provide cooling water. Also, it is easier to persuade local communities to accept a nuclear facility if half the affected people live across a frontier and so have no say in the matter.[27]

These transboundary disputes over nuclear policies and plants include the Danish government, which is opposed to nuclear power, asking Sweden to close a reactor close to Copenhagen; demonstrators from Luxembourg and West Germany crossing the border to protest a French nuclear plant that they consider unsafe; the Austrian government asking West Germany to stop building a nuclear reprocessing facility, accompanied by large Austrian demonstrations at a makeshift chapel near the plant; and Irish citizens and politicians asking for the closure of a British reprocessing plant that has allegedly dumped large amounts of nuclear waste illegally into the Irish Sea.[28] Beyond these conflicts over nuclear power generation, clashes have occurred over the movement of commodities contaminated by Chernobyl radiation: for example, Italian border patrols detained thirty-two freight cars loaded with cattle, sheep, and horses from Austria and Poland before forcing them to return due to abnormally high radiation levels;[29] and an attempt by West German businessmen to ship 4800 tons of Chernobyl-contaminated whey powder to Egypt met with rejection.[30]

The conflict within Western European nations occurred largely because many Europeans "lost faith in government officials whom they believe deliberately understated the health threat from the accident"; public criticism and confusion grew as a governmental "mixture of paternalism, bureaucratic incompetence, and lack of preparation turned out to be a chaotic brew."[31] For example, the Italian government left it to local authorities and citizens' groups to monitor radiation levels; the British and French governments falsely announced that their countries were spared;[32] and the Greek government provided such incomplete information that a frustrated private group of doctors, lawyers, and scientists formed to measure food radiation themselves.[33]

CONSEQUENCES OF THE CHERNOBYL DISASTER

The principal impact of the Chernobyl accident was to increase dramatically (if not permanently) the strength and determination of antinuclear sentiment and the willingness of nations using nuclear energy to open up their programs and plants for international scrutiny. This effect occurred to varying degrees within the Soviet Union, Western Europe, and the international community as a whole.

The Soviet Union suffered severe economic and political setbacks as a result of Chernobyl. The estimated three-billion-dollar cleanup bill, mainly for equipment and conscripted workers brought in from throughout the nation, did not include the costs of using oil that would have been exported to generate substitute electricity, of medical monitoring to be conducted for decades to come, or of major losses to Soviet agriculture.[34] The Soviet Union lost two thousand megawatts of power indefinitely, and its energy expansion program has been set back.[35] The Russians have decided to implement modifications of all of their graphite-moderated reactors and not to construct any more reactors of this type, although they cannot afford to abandon any of their existing nuclear facilities.[36] Politically, the Soviets experienced increased tensions in arms-control talks with the United States immediately following the disaster;[37] Mikhail Gorbachev's image of "decisiveness and openness" was tarnished internationally; and Chernobyl seems "certain to intensify debate about socioeconomic reforms and information policy" within Soviet decision-making bodies.[38]

The consequences of the Chernobyl disaster on Western Europe involved the most acute growth of public opposition to nuclear energy. Flavin notes that most of the 1970s anti-nuclear movement had peaked and was in decline until Chernobyl, after which there was "a rebirth of large antinuclear demonstrations throughout Western Europe," to the point where more than two-thirds of the population in most European nations (with France a distinct exception) opposed nuclear power.[39] In Switzerland and Italy, citizens were scheduled to vote on whether to curtail nuclear power.[40] In Sweden, anti-nuclear public opinion hardened, and the government intends to eliminate nuclear power by 2010.[41] In the United Kingdom, the Labour Party called for a ten-year phaseout of nuclear power.[42] In

Austria and Greece, governments decided not to begin using nuclear power.[43] While some have begun to fear "the possible abandonment of nuclear power" in Western Europe,[44] Fischer points out that, because Western Europe is already drawing 30 percent of its electricity from nuclear plants, and a switch to coal would take several decades and harm the environment, the leaders of the major Western powers at a meeting in Tokyo in May 1986 reaffirmed their support of "properly managed" nuclear power.[45]

Finally, the Chernobyl accident resulted in the international community taking some cooperative steps to manage nuclear energy. Controversy exists about whether existing international organizations were effective during the disaster: some argue that "the international community showed that it was not even remotely prepared for such an emergency," as the International Atomic Energy Agency (IAEA) provided "no food or health recommendations," the World Health Organization (WHO) waited a week and then issued only broad warnings, and the European Economic Community (EEC) waited for three weeks to deal with food imports and crop radiation levels and then "the standards were weak and clouded by political compromises."[46] Others contend that "international organizations interested in nuclear safety acted quickly to assess the accident's implications."[47] Although the more skeptical interpretation seems more realistic in the period immediately following the disaster, Chernobyl demonstrated that "one of the few beneficial effects of the tragedy" was "a renewed international willingness to cooperate on nuclear safety."[48] On May 21, 1986, the IAEA's thirty-five member board agreed (with strong Soviet support) to a plan to (1) draft a binding agreement for early notification and full information about nuclear accidents with possible transboundary effects; (2) draft a binding agreement to coordinate emergency responses to such accidents; (3) hold a full-scale international inquest on the accident; (4) improve nuclear safety cooperation; and (5) convene a conference on the full range of nuclear safety issues.[49] On May 9, 1986, a committee of the Nuclear Energy Agency of the Organization for Economic Cooperation and Development (OECD) met and decided to study action needed to strengthen OECD cooperation in case of severe nuclear accidents. In addition, the Common Market nuclear energy agency—Euratom—has sought to bring uniformity into its

9

Evaluation of Case Study Patterns

The five international resource conflicts described in the preceding chapters contain some tentative but intriguing patterns about the causes and consequences of such clashes. This chapter evaluates the theoretical framework presented in Chapter 2 in light of these case studies and then analyzes other commonalities evident in the cases. In the process, a number of sweeping intuitive assumptions about resource conflict are called into question or more narrowly qualified.

EVALUATION OF THEORETICAL CONTENTIONS IN LIGHT OF CASE FINDINGS

A juxtaposition of the results of the case studies to the general theoretical propositions about the causes and consequences of conflict over the world's resources yields generally positive but mixed results. Of course, not every theoretical contention presented earlier is confronted by evidence from the case studies. Figure 11 summarizes the principal findings from the cases on resource scarcity, national development, resource inequality, resource interdependence, resource time lags, and conflict impacts. This section analyzes the similarities and differences between theory and evidence in accordance with the sequence of elements in the framework, as depicted in Figure 3.

Figure 11
Patterns in Resource Conflict Cases

	WHALING CONFRONTATION	*OIL CRISIS*	*GRAIN COERCION*	*STRATEGIC MINERALS THREAT*	*CHERNOBYL NUCLEAR DISASTER*
SCARCITY	Predictable, Slow+Small	Unexpected, Rapid+Massive	Medium Rates	Predictable, Slow+Small	Unexpected, Rapid+Massive
	Intended+ Direct	Intended+ Direct	Intended+ Direct	Feared+ Indirect	Unintended+ Indirect
	Real+ Global	Contrived+ Global	Contrived+ Local	Contrived+ Local	Real+ Local
supply	Decreasing+ Ecological	Decreasing+ Political	Decreasing+ Political	Decreasing+ Political	Decreasing+ Ecological
demand	Constant+ Psychological	Increasing+ Technological	Increasing+ Ecological	Increasing+ Technological	Constant+ Ecological
DEVELOP-MENT *initiator*	Rising Expectations Constant Achievements	Rising Expectations Constant Achievements	Rising Expectations Constant Achievements	Constant Expectations Falling Achievements	Constant Expectations Falling Achievements
victim	Constant Expectations Falling Achievements	Rising Expectations Falling Achievements	Constant Expectations Falling Achievements	Constant Expectations Falling Achievements	Constant Expectations Falling Achievements

INEQUALITY	Decreasing in Consumption (Victim Greater)	Increasing in Control (Initiator Greater)	Constant in Control (Initiator Greater)	Constant in Control (Initiator Greater)	Increasing in Damage (Initiator Greater)
	Intolerance of Disparity	Intolerance of Disparity	Tolerance of Disparity	Tolerance of Disparity	Tolerance of Disparity
INTERDE-PENDENCE	Continuing Reciprocal	Growing Skewed	Growing Reciprocal	Continuing Skewed	Continuing Reciprocal
TIME LAGS	Short Ecological Medium Psychological Long Political	Short Psychological Long Technological	Medium Ecological Long Political	Long Psychological Medium Political	Short Ecological+ Psychological Long Political+ Technological
IMPACTS:					
scarcity	Decreased	Decreased	Decreased	Decreased	Decreased
develop-ment	Initiator-- Constant Victim-- Constant	Initiator-- Increased Victim-- Decreased	Initiator-- Constant Victim-- Constant	Initiator-- Constant Victim-- Constant	Initiator-- Decreased Victim-- Decreased
inequality	Decreased	Increased	Constant	Constant	Constant
interde-pendence	Increased	Decreased	Decreased	Decreased	Increased

Resource Scarcity

With regard to the resource disruption that began each conflict, there were differences among the five cases in the speed, size, expectedness, intentionality, and directness of this break from prevailing resource relationships. The oil crisis and the Chernobyl nuclear disaster were unexpected (Chernobyl much more so), rapid, and massive disruptions; the whaling confrontation and strategic minerals threat (to the extent the threat has been realized) have been predictable, slow, and small (incremental) disruptions; and the grain coercion was in between in size, speed, and expectedness. The whaling, oil, and grain cases were intended and direct disruptions, implemented through either governmental political sanctions or transnational (corporation or conservationist) activities restricting access to a desired resource. On the other hand, the strategic minerals and Chernobyl cases exhibited clouded intentionality and were more indirect disruptions: for strategic minerals, the threat may be at least as much a product of American fears as Soviet intentions, and the Soviets would have to work through the governments of mineral-exporting states to deny these resources to the United States; and with Chernobyl, the nuclear accident was clearly unintended, and the radioactive fallout first had to spread from the Soviet Union and contaminate the Western European environment before interfering with the quality of life in that area.

The trends in resource scarcity at the outset of the five conflicts flow smoothly from these patterns in resource disruption. In the whaling confrontation, the scarcity was real and global, reflecting natural limits in the ecological supply of whales worldwide, combined with an intense psychological addiction to their consumption, particularly in Japan. In the oil crisis, the scarcity was contrived and global, reflecting a political bottleneck in the global supply of petroleum and a technological (industrial) dependence by the West on this resource. In the grain and strategic minerals cases, the scarcity seemed contrived and local, reflecting political minerals for the United States, combined with an ecological Russian demand for grain to feed hungry people and a technological American demand for strategic minerals to satisfy military/industrial needs. In the Chernobyl disaster, the scarcity of uncontaminated resources was real and local, reflecting severe ecological impediments to traditional supply and demand relationships needed to sustain the European and Russian populations. Furthermore, in the whaling and Chernobyl cases, nations displayed a constant demand but faced a decreasing supply of needed/desired resources; while in the oil, grain, and strategic minerals cases, demand escalated, reflecting decreased self-sufficiency, and supply plummeted.

The theoretical propositions cited in Chapter 2 pertaining to resource scarcity indicate that national frustration and international resource conflict would be more likely or severe when (1) the disruption was unexpected, rapid, and massive rather than predictable, slow, and small; (2) the scarcity was contrived, based on intended psychological, technological, or political disequilibrium, rather than real, based on unintended or natural ecological disequilibrium (even though such ecological disequilibrium often reflects a more severe scarcity); and (3) scarcity encompasses the inelastic combination of increasing demand and decreasing supply. Only the oil crisis—the most severe of the resource conflicts—fits all three conditions, and the theoretical conditions do appear reasonable consistent with the case findings. But the cases help to qualify the sweeping nature of these propositions: the Chernobyl nuclear disaster shows that even the most unexpected, rapid, and massive resource disruption can have the tensions it generates inhibited and diffused somewhat by the realization that the incident was not intended and that it reflected real ecological scarcity of uncontaminated resources, with the worst effects experienced by the initiator itself (the Soviet Union); and the grain and strategic minerals cases show that even resented contrived scarcity and the most disequilibrated economic situation reflecting escalating demand and plummeting supply can have their conflict potential somewhat diminished by the resource disruption being predictable, slow, and small.

Development

Consideration of the transformation of national development resulting from resource scarcity requires separate examination of the productivity of initiators and victims in the five conflict cases. The initiators, who in all the conflict cases except for the oil crisis were developed rather than developing nations, at the outset of the conflicts experienced rising resource expectations and constant resource achievements in the whaling confrontation, oil crisis, and grain coercion; and constant expectations and falling achievements in the strategic minerals threat and the Chernobyl nuclear disaster. The victims, who were entirely developed nations, experienced constant resource expectations and falling resource achievements in the whaling, grain, strategic minerals, and Chernobyl cases; and rising expectations and falling achievements in the oil crisis.

More specifically, conflict initiators with the rising-expectation/constant-achievement combination hoped to use their growing leverage to break themselves out of a stagnant predicament: the IWC members and transnational conservationists wanted finally to be able to use their growing political clout to force Japan and the Soviet Union to curtail their whaling operations; and the OAPEC members and the United States wanted to use their growing oil and grain power (respectively) to force major economic concessions out of the Western nations and the Soviet Union. Conflict initiators with the constant-expectation/falling-achievement combination planned simply to maintain existing development levels but found themselves unable to do so: the Soviet Union has had its resource self-sufficiency somewhat stymied by its increasing dependence on foreign sources of strategic minerals, and its energy plans stymied by the destruction of a nuclear power plant and subsequent environmental degradation. Turning to conflict victims, those with the constant-expectation/falling-achievement combination were, like the initiators, simply attempting to maintain their productivity pattern but were unsuccessful: Japan and the Soviet Union attempted to continue whaling but were impeded by political sanctions and physical interference; the Soviet Union attempted to feed its people but was impeded by bad harvests and American embargoes; the United States has attempted to continue developing its military arsenal but fears resource denial by the Soviets; and Western Europe attempted to continue "business-as-usual" but was impeded by the fallout from Chernobyl. Finally, conflict victims with the rising-expectation/falling-achievements combination were planning to expand their national productivity when their economies were thrust in the opposite direction: the hopes for an economic boom by Japan, the United States, and Western Europe were dashed and replaced by recession due to the price rises emanating from the Arab oil embargo.

The theoretical contentions about the relationship between development and resource conflict indicate that such conflict would be more likely and more severe when declining development occurs, reflecting a growing societal gap between development expectations and development achievements with respect to the resource in question. The initiators and victims in all five of the conflicts experienced this growing gap, which is at its worst in the oil crisis (the most severe conflict), where victims' expectations and achievements both change in opposite directions. Thus the theory, which admittedly is broad and vague, is reasonably well substantiated by the cases. It is worth noting that all of the victims experienced falling development achievements with respect to the given resource soon after the resource disruption and the onset of the conflict; but such productivity declines could be attributed to natural forces as much as to the success of initiators' efforts. For example, declining whaling resulted from the ecological scarcity of the resource almost as much as from IWC and conservationist sanctions, and Soviet grain shortages resulted from climate fluctuations more than from any American grain sanctions.

Resource Inequality

The issue of resource inequality between initiators and victims in the five conflict cases was as much an issue of existing resource disparities as that of transformed distribution with respect to the resource in question. Though all of the cases except the oil crisis reflected roughly comparable equality in the overall resource strength of initiators as compared with victims, focusing on the particular resource involved in each dispute reveals significant discrepancies at the outset of the conflicts. In the whaling confrontation, although no nation truly owns or controls whales, the victims consumed significantly more of the resource than the initiators; in the oil crisis, grain coercion, and strategic minerals threat, the initiators possessed significantly more of the resources than the victims; and in the Chernobyl nuclear disaster, the initiator experienced greater damage from nuclear pollution than did the victims. Furthermore, in the whaling, oil, and Chernobyl cases, the inequality during the resource conflict was a major change from the inequality prior to the resource disruption: the whaling nations generally consumed less of the resource, reducing inequality with nonwhaling IWC nations; OAPEC members gained more control over multinational oil companies, increasing inequality in energy clout with the West; and the Soviet Union experienced greater energy and environmental setbacks, increasing inequality in damage with Western Europe. Intolerable discrepancies in resource consumption or living standards were present only in the whaling confrontation, where conservationists resented whaling by Japan and the Soviet Union, and the oil crisis, where Arab states resented the "fat cat" lifestyles in the West; in the other cases, little resentment was evident between initiators and victims about the level of resource use during the conflicts.

Theories concerning resource inequality suggest that frustration and international resource conflict seem most likely or severe when (1) significant inequalities exist in dimensions deemed important; (2) these inequalities are much larger than previous ones; and (3) these inequalities reflect discrepancies which are intolerable to relevant parties. These theoretical contentions do receive reasonable support from the case findings: the most severe case, the oil crisis, satisfies all three conditions, and all of the cases contain significant resource inequalities between initiators and victims. But only the oil and Chernobyl cases involved inequality significantly higher than the previous status quo, and only the whaling and oil cases evidenced deep intolerance of discrepancies. One explanation of this last finding is that resentment of inequality would seem to be most expectable in North-South relationships, where there is a long legacy of distrust and exploitation, and the oil crisis is alone among the cases in reflecting this pattern of resource conflict (which is the second conflict type presented in Chapter 1).

Resource Interdependence

As with inequality, the pattern of resource interdependence in the five resource conflicts involves several facts of the linkages between initiators and victims. Given the frequency of East-West confrontations among this study's cases, it is perhaps not surprising that the overall resource interdependence between initiators and victims was high only in the whaling confrontation (for Japan) and the oil crisis. However, when concentrating specifically on the resources involved in the disruptions, interdependence was high in all of the cases: Japan and the Soviet Union depended on the United States for fishing rights in American territorial waters, and several anti-whaling nations depended on the Soviets for raw materials and on the Japanese for technology and manufactured goods; the West depended on OAPEC for oil, and OAPEC depended on the West for technology and (to some extent) military security; the Soviet Union depended on the United States for grain, and the United States has depended on the Soviet Union as an outlet for its agricultural surpluses and as a source for strategic minerals (and in 1975 was hoping to become dependent on the Soviets for low-cost oil); and the uncontrollable spillover effects of nuclear pollution have created interdependencies between the Soviet bloc and Western Europe in nuclear energy usage. But this interdependence specific to the conflict-producing resource had been recently heightened only in the oil and grain cases. Furthermore, the interdependence was roughly reciprocal in the whaling, grain, and Chernobyl cases, and skewed in favor of the initiators in the oil and strategic minerals cases. Finally, all five of the resource conflicts involved initiator-victim relationships characterized at least to some degree by antagonism and misunderstanding.

According to theories of resource interdependence, international resource conflict is most likely and most severe when the following conditions are present: (1) high initiator-victim interdependence; (2) a recent increase in this interdependence over past levels; (3) skewed rather than reciprocal interdependence; and (4) traditional initiator-victim antagonisms and misunderstandings. The case findings reasonably support these propositions, as the first and last conditions are confirmed in all five cases, and once again the most severe case on oil is alone in fulfilling all four requirements. These findings do, however, help to limit the scope of broad theoretical claims: the grain case shows that recently heightened interdependence (especially true in the 1975 rather than 1980–81 embargo) between the United States and the Soviet Union may have its tendency to produce conflict somewhat hampered by the reciprocity of the interdependence and the availability of alternative grain suppliers; and the strategic minerals case suggests that highly skewed interdependence may have its conflict tendencies tempered by the absence of recent change increases in interdependence levels (although American foreign dependence on strategic minerals had grown, its dependence specifically on the Soviet Union for these minerals had not done so).

Resource Time Lags

Time lags were central in both the causes and consequences of the five resource conflicts. The whaling confrontation had a short ecological time lag, as the whale population was near extinction; a medium psychological time lag, as it took some time for the public and national governments (especially of whaling nations) to become aware of the ramifications of the whaling issue; and a long political time lag—perhaps the longest of any of the cases—because of the resistance to whaling regulation and the loopholes in restrictions. The oil crisis had a short psychological time lag, as masses and elites in Western nations immediately felt the brunt of the embargo through price rises and, to a lesser extent, supply shortages; and a long technological time lag, as attempts to find substitutes for oil or new reserves of oil in response to scarcity were slow and only partially successful. The grain coercion involved a medium ecological time lag between the timing of poor Soviet harvests and the shortages to the Russian people; and a long political time lag between American imposition of the grain embargoes, any impacts on the Soviet economy, and the lifting of the embargoes. The strategic minerals threat has had a long psychological time lag, as most of the American public and many foreign-policy makers are still unaware of the issue; and a medium political time lag, as even the supportive Reagan administration took considerable time before generating policies to cope with the threat. Finally, the Chernobyl nuclear disaster had short ecological and psychological time lags, as the environment was quickly contaminated and people quickly became aware of the problem, but long technological and political time lags, as considerable sluggishness was evident in developing technological assessments of and solutions to the radioactive fallout and in coordinating political efforts to manage the catastrophe.

What little theory exists concerning time lags implies that resource conflict should be most likely or severe when ecological time lags are short and psychological, technological, and political time lags are long. This set of assumptions is clearly not generally supported by the case findings, as only whaling (one of the more mild conflicts) and Chernobyl approximated these conditions. A short ecological time lag was not present in the other cases because the resource disruption revolved around contrived rather than real scarcity, eliminating the precipitant of environmental degradation or severe ecologically based global shortages. This finding serves to reinforce the point that modern resource battles need not have their roots in ecological crisis.

Conflict Consequences

Turning directly to the impacts of the resource conflicts, significant changes emerged in scarcity, development, inequality, and interdependence in the five cases after the disputes flared up. Resource scarcity decreased in all cases as a result of the conflicts, due to a reduction of demand for the resources in question, diversification of resources used or supply sources of the given resource, stockpiling of the resource, and/or reparation of the environment. Development largely remained unaffected in initiating and victim nations, except that the oil crisis increased development among OAPEC's members and decreased it among Western nations, and that the Chernobyl nuclear disaster produced decreased development in both the Soviet Union and Western Europe. Resource inequality between initiators and victims also was mostly unchanged except in the whaling and oil cases, in which the whaling nations' catch was curtailed and Arab states' oil control was consolidated (at least until the 1980s) as a result of the resource conflicts. Lastly, resource interdependence between initiators and victims increased as a result of the whaling and Chernobyl cases, in which international cooperation became more explicit and enforced; and decreased in the wake of the oil, grain, and strategic minerals cases, in which victims chose to reduce their reliance on volatile and politically motivated suppliers.

The extremely sparse theoretical threads discussing the consequences of resource conflict suggest its impacts are growing scarcity, inequality, and interdependence, with development unaffected in the aggregate (though increasing in some sectors and declining in others). Clearly none of these suppositions are fully supported by the case findings, which present opposite outcomes for scarcity and more mixed results for development, inequality, and interdependence. The weakness of the theory, much of which was admittedly designed to deal with the impact of violent military conflict, is the primary culprit here, and generally seems to understate the adaptive abilities of the conflicting parties to take steps after a clash to reduce relevant resource tensions.

Synthesis

Examining the overall correspondence of the theoretical framework to the case findings reveals that contentions concerning the four principal elements of the framework—scarcity, development, inequality, and interdependence—receive reasonably strong support, while those concerning time lags and conflict consequences receive little substantiation from this evidence. Moreover, a couple of interesting patterns emerge within/among the elements of this framework: (1) those conflicts reflecting real scarcity are characterized by constant demand trends, while those reflecting contrived scarcity are characterized by increasing demand, suggesting that nation-manipulated resource bottlenecks may be more likely to be applied just when the squeeze is tightest; and (2) the intended and direct resource disruptions are those in which initiators had rising expectations and constant achievements (rather than constant expectations and falling achievements), suggesting that nations may be more likely to initiate resource conflict out of confidence rather than desperation.

OTHER RESOURCE CONFLICT PATTERNS

Beyond the relationships highlighted by the theoretical framework, the case studies reveal a number of other important aspects of the causes and consequences of conflicts over the world's resources. While these additional patterns may not be as rooted in articulated theory, some seem especially critical in light of the policy prescriptions to be developed in Chapter 10.

Turning first to the causes of conflict, the cases demonstrate that the conflicts were launched by producers/exporters/caretakers of the resource in question, with consumers/importers as the targets, further reinforcing (along with earlier patterns) that resource vulnerability is more likely to be associated with victims and resource strength with initiators. Each conflict involved a resource deemed by the victims to be vital to their survival in a different way, and this sense of necessity helped to intensify tensions: the Japanese felt that whale meat was vital to their diet (not to mention the conservationist view that whales are vital to human survival through their involvement in the ecosystem's production of oxygen); the West felt that oil was essential for commercial, residential, and transportation needs; the Soviet Union required grain to feed its people; the United States had defense and industrial systems that it sensed would collapse without strategic minerals; and Western Europe (not to mention the Soviet Union) could not survive in an environment heavily contaminated from nuclear fallout. In each of the cases, the supply of scarcity seemed more important in causing conflict than the demand roots, as without the disrupted/diminishing supply of the desired resources the demand fluctuations would not have been sufficient to create critical tensions. The weakness of relevant international organizations also contributed to the onset of conflict in each case: the IWC had not been able in the past to manage whale stocks effectively; the OECD engaged in little contingency planning for the oil embargo and was somewhat ineffective once it began; the World Food Conference in Rome had been unable to achieve significant agreements about international grain exchange before the grain embargoes; the unsatisfactory state of the UN Law of the Sea Treaty exacerbated strategic minerals tensions; and the IAEA seems to have come up a bit short in undertaking needed actions quickly to restrain disputes arising from Chernobyl. Finally, each case had its intergovernmental tensions compounded by international clashes between national governments and transnational groups: conservationist groups pressured whaling governments (the victims) to stop their hunting and non-whaling governments (the initiators) to support stronger sanctions; oil companies fought pressures from oil-importing governments (the victims), while succumbing to the influence of oil-exporting governments (the initiators); grain companies helped to thwart both American embargoes despite the best efforts of the American government (the initiator) to prevent having its actions undermined; mining companies may fulfill ominous warnings and not continue to secure strategic minerals needed by the American government (the victim), and anti-apartheid groups may interfere with relations with mineral-exporting South Africa; and anti-nuclear groups pressured Western European governments (the victims) to alter official pro-nuclear positions.

Moving to the consequences of conflict, in direct relation to the last causal pattern, the cases generally created a loss of faith by the informed public about the ability of their own governments to manage the ongoing resource situation effectively, which in turn reduced mass support for government initiatives to cope with the disruptions. Moreover, those cases that involved multiple initiators or victims found tensions across these nations heightened: strains widened among IWC members as the whaling confrontation progressed, among OAPEC members and among Japan, the United States, and Western Europe after the oil crisis, and among affected Western European nations after the Chernobyl nuclear disaster. While some increased international cooperation did emerge as a result of most of the conflicts, the significant policy coordination appeared to be mostly temporary. However, in every case the victims responded to the conflict by taking steps to reduce their use or external reliance on the resource in question: Japan and the Soviet Union reduced their whaling, the West used less oil and imported less from OAPEC, the Soviet Union relied less on American grain, the United States tried to reduce its external dependence on strategic minerals, and Western Europe scaled down some nuclear power programs.

More generally, these international resource conflicts highlighted the tension between the goals/values of resource security and environmental protection. In the three cases emphasizing resource security—oil, grain, and strategic minerals—preservation of the natural environment not only took a back seat but was actually hurt by the security focus, as victims of resource disruptions scrambled to expand domestic sources of supply, no matter what the ecological costs. This priority is at least partially a function of the overlap between national resource security concerns and the East-West Communist-capitalist cleavage.

Given the prevailing priority on national resource security over environmental protection, it is perhaps not surprising that the conflicts which contributed the most to cooperative international resource management were the whaling and Chernobyl cases, which centered on environmental protection and common-pool issues that could not be confined to any national setting. The unilateral use of resource weapons for purposes of enhancing national security can achieve at least temporary gains for the initiators, as the oil crisis demonstrates, but the costs are large: unforeseen side-effects on non-target nations, potential backfire effects on the initiators, internal bickering, global resentment, and loss of control due to the complex and multifaceted international resource exchange system. The seductive attraction of linking political ends to resource strengths will, nonetheless, continue.

10

Policy Implications

The preceding discussion of case study patterns is conducive to the development of policies for minimizing conflict over the world's resources. This chapter places these prescriptions in the context of the specific solutions suggested with respect to each of the five conflicts, considers the important policy obstacle of prevailing attitudes toward resources, and then reaches some overarching conclusions about the prognosis for resource conflict. While this study has attempted throughout to avoid taking value positions, the normative basis of the policy implications presented here is simply the desirability of conflict prevention/reduction.

The need for meaningful policy prescription appears to be particularly acute in dealing with global resource tensions because so much of the advice that has emerged has been so vague and idealistic that policy makers have perceived it as irrelevant in the short-term and as impossible to implement. For example, Westing forwards the following suggestions as means "to eliminate competition over . . . resources as a source of international conflict:"

> (a) for the living natural resources (whether national, shared or extra-territorial), an inviolate balance must be established between harvesting and natural regeneration; (b) for all natural resources that overlap national boundaries, formal mechanisms under international auspices must be estab-

> lished for equitable sharing; and (c) for all extra-territorial resources it must be accepted that they constitute a common heritage of humankind, and bodies of international law must be established accordingly that provide for their equitable sharing for the benefit of all.[1]

This kind of long-range wish list, if left in isolation, can often lead to stagnation in international environmental action.

PRESCRIPTIONS FOR MINIMIZING INTERNATIONAL RESOURCE CONFLICT

While the five conficts discussed in this study provide numerous specific policy recommendations, many of these are similar to actions already taken in the wake of the clashes or are aimed as much at resolving resource shortages as at minimizing international resource conflict. Nonetheless, these suggestions appear to be quite useful as a starting point, and four themes run through these specific proposals: (1) improving the collection and dissemination of resource information; (2) strengthening international cooperation and existing international organizations; (3) encouraging conservation (reduced consumption) of the resource in question; and (4) fostering the development and use of substitutes for the given resource. These solutions include short-term and long-term, conservative and radical, and nationalist and internationalist ideas.

Regarding the whaling confrontation, some of the more short-term and conservative prescriptions are (1) obtaining better scientific information on the size of existing whale stocks, perhaps through joint American-Japanese reserch; (2) extending and enlarging the power of the IWC; (3) relying more on plentiful substitutes for whale products, such as soybean cakes for whale meat and oil from the jojoba plant for whale oil; and (4) increasing communication between groups/nations for and against whaling so as to promote mutual understanding.[2] Some more radical long-term approaches are (1) initiating a new United Nations whaling convention; (2) having American-financed benefits to whaling nations to compensate for lost revenues due to curtailed operations; and (3) creating a global agency empowered to sell a restricted number of whale-hunting licenses to nations.[3]

Regarding the oil crisis, short-term nationalist policy recommendations include (1) stockpiling oil in case of emergency; (2) reducing dependence on OAPEC oil, and (3) increasing energy self-sufficiency.[4] More long-term internationalist approaches are (1) strengthened IEA and OECD cooperation on energy issues; (2) increased information exchange among industrialized nations about energy consumption and production; (3) more attention to energy substitutes for oil, including "soft" energy sources, which need to be wrested from the control of the major oil companies; (4) further exploration and development of new oil sources; (5) encouraging energy conservation; and (6) heightened efficiency in energy use.[5]

Regarding the grain coercion, from a short-term nationalist position the American government has a number of "unpalatable" options available for dealing with its excess grain: (1) buying the grain and storing massive amounts of it until the world market revives; (2) buying the grain and giving it to the world's poor and hungry, who cannot afford world market prices; and, most likely, (3) attempting aggressively to win the remainder of the world's market with various kinds of export subsidies supporting the American market share.[6] From a more long-term internationalist perspective, policy alternatives include (1) multilateral guarantees for food aid during shortages, including the use of substitutes for a particular scarce food resource; (2) the creation of an International Food Bank; (3) improvement in the free flow of agricultural information; and (4) a reduction in the international demand for grain, especially imports from nations attaching political strings to its provision.[7]

Regarding the strategic minerals threat, most of the attention is on policy recommendations narrowly oriented toward maximizing U.S. security interests: (1) improving the content and quality of the national stockpile of strategic minerals, including the possibility of privately held stockpiles; (2) encouraging domestic extraction and production of these minerals, including tax incentives and increased research-and-development funds; (3) fostering greater foreign investment by multinational mining companies, including government investment guarantees, risk insurance, and loans; (4) easing environmental regulations and opening up federally protected lands to minerals exploration; (5) promoting Defense Department efforts to develop new manufacturing technologies and synthetic materials that reduce external dependence on strategic minerals; and (6) using diplomacy and economic and military aid to prevent international conflicts in mineral-producing regions of the world, including increased producer-consumer agreements and pressures to reduce communist intrusion in these regions.[8] But some suggested policies have more direct internationalist overtones: (1) coordination among Western allies in stockpiling and emergency response planning, creating joint research-and-development programs, and providing assistance to mineral-producing states; (2) improving international methods for the collection and use of mineral data; (3) increasing research on mining minerals from continental shelf areas, the seabed floor, and outer space; (4) conservation and recycling of these minerals; and (5) development and use of substitutes for scarce minerals.[9]

Finally, regarding the Chernobyl nuclear disaster, most of the prescriptions are short-term, conservative, and internationalist: (1) a stronger IAEA; (2) binding minimum safety standards for nuclear power plants; (3) punishment administered on states violating nuclear guidelines; (4) greater international scrutiny of national nuclear programs, including expanded on-site safety inspection systems; and (5) wider availability of information on nuclear energy generation and usage, including continuous global evaluation of radiation levels and their effects.[10] But a few more radical, long-term ideas also emerge in this context: (1) the technological development of "inherently safe" nuclear reactors, which do not depend on mechanical or human intervention, and (2) a significant reduction in the overall global use of nuclear power and reliance instead of other energy sources.[11]

The four prescriptions mentioned in all of these cases—information gathering, international cooperation, resource conservation, and resource substitutes—fit in well in addressing the tensions causing resource conflict identified by the theoretical framework. Encouraging conservation and developing substitutes link up with decreasing resource scarcity, the first through reducing demand and the second through expanding supply. Improving information collection and strengthening international organizations tie into improving resource interdependence, as these policies generally serve to increase the sense of both reciprocity and continuity in interdependent relationships. Moreover, the enhanced information can help to shorten frustrating psychological, technological, and political time lags. It appears that decreasing scarcity is a more basic element in minimizing resource conflict and is frequently a prerequisite to improving resource interdependence.

However, the prescriptions emerging from the cases do not seem in themselves sufficient to cope with the full spectrum of tensions precipitating conflict over the world's resources. Though many of the causes of such conflict seem relatively immune to practical, implementable policy recommendations—such as the speed and size of resource disruptions, the falling development achievements of victims, and the unequal control of key resources—there are at least a few additional possibilities. First, it seems essential to reduce the conviction on the part of any nation that a particular natural resource is absolutely vital. Modernization has often meant standardization not only of the technology of production but also of the resources upon which the technology draws: for nonrenewable resources, a critical need exists to diversify the resource base of industry, not just through discovering resource substitutes but by developing flexible technologies which can rely on a variety of different resources; and for renewable resources, addiction to one particular variety—such as a special kind of food—cannot be tolerated internationally without running the risk of raising resource tensions. Second, it appears important to bridge the gap between the public and private sectors on resource issues, for national governments' inability to coordinate their policies with multinational resource companies and transnational conservationist groups can create paralysis or mutually undercutting actions. There need to be more formal channels of communication and consultation between governmental and nongovernmental groups concerned with resource issues, particularly in nations (such as the United States) where a tradition of adversarial relationships exists.[12] Third, it seems desirable to have more third-party mediation used to deal with imminent or ongoing conflicts. While this study's five cases contain little effective international mediation, this technique has been specifically suggested for and used in international resource conflicts, and would seem to have the greatest chances of success when a competent intermediary (combining ecological, technological, and political expertise) is available and when the parties to a dispute view each other as legitimate and roughly equal.[13]

OBSTACLE OF PREVAILING ATTITUDES TOWARD RESOURCES

Regardless of the policy approach taken to minimize international resource conflict, perhaps the major stumbling block threatening chances for success is the dominant manner in which individuals, groups, and nations view natural resources. As the case studies demonstrate, many see these resources simply as indefinitely available tools to satisfy selfish craving or to achieve more grandiose political ends. There is not only little understanding of the environment and of ecological limits on resource production, but also scant comprehension of tradeoffs involved in resource manipulation—between resource security and environmental protection, resource use and pollution, short-term and long-term resource issues, and resource exchange and resource vulnerability. The gap in understanding among foreign-policy makers has generally led them in their resource decisions not toward caution, restraint, and open-minded questioning but rather toward recklessness, daring, and close-minded certainty. Altering these counterproductive and conflict-producing perceptions is one of the principal prerequisites to sound management of resource conflicts.

More specifically, Figure 12 depicts the traditional mode of operation of the international system, revealing a selfish hierarchy of resource growth/distribution concerns. The dominant governmental concern is for growth over distribution and for the needs of one's own nation in the present over needs of others and future needs. The principal regulatory mechanism of free-market economies—the price system—may not take sufficiently into account long-run future needs,[14] and social taboos may interfere with balancing today's demand against tomorrow's supply.[15] Under scarcity, this hierarchy would be further skewed toward current selfish concerns because of the inability to satisfy demand. Catering to current internal resource-growth needs may infuriate other nations, and catering to current external resource distribution may infuriate one's own citizenry; overwhelming concern for the future resource needs of the world as a whole would seem to infuriate everyone. The predominant emphasis on the growth needs of the present generation within one's own society seems not only to be the most selfish focus but also the one least compatible with a steady-state system, most obsessed with

Figure 12
Hierarchy of Resource Growth/Distribution Concerns

GROWTH	*Selfish*	Needs/Desires of Present Generation	1	3
	Altruistic	Needs/Desires of Future Generations	2	4
		DISTRIBUTION	Needs/Desires of One's Own Nation *Selfish*	Needs/Desires of Other Nations *Altruistic*

RANKING:
1=Top Priority
4=Bottom Priority

wants over needs and with independence over interdependence, and most likely to generate pervasive international resource conflict. While some contend that present resource use is not at the expense of that of future generations, this position rests on extremely optimistic assumptions about the role of the price system and technological innovation.[16] The generally greater governmental enthusiasm in the conflict cases revolving around resource security—oil, grain, and strategic minerals—for nationalist rather than internationalist policy recommendations testifies both to the intrusiveness of this hierarchy of concerns and the need to modify it in order to minimize resource conflict.

CONCLUSION

The recent history of conflict over the world's resources, briefly chronicled in this book, has not provided the basis for much optimism about the immediate future. The absence of learning from past experience in the resource arena has been noteworthy, as each new encounter seems to be approached in something of a conceptual vacuum. Of all the cases, the grain coercion demonstrated this sobering reality most graphically, as within a five-year span the same nation used the same resource weapon against the same target with the same unfortunate conflict consequences. Despite the myriad of conflict-minimizing policy options available, the sources of resource conflict appear to be too powerful and pervasive to permit a great reduction in the frequency of these disputes, and their product is usually deep-seated hostility that does not vanish easily. Resource issues, which in some ways are more concrete and basic than other international concerns, can provide a kind of barometer of the sophistication and advancement of relationships among global actors, and according to that standard recent international performance cannot be given a high rating. For better or worse, raw geopolitical concerns still dominate the world today and seem somewhat out of control.

This book has attempted to lay the groundwork for a systematic comparative and integrative treatment of conflicts over the world's resources. After developing a theoretical framework focusing on ecological, psychological, technological, and political aspects of scarcity, development, inequality, and interdependence, the study evaluated emerging conceptual propositions through scrutiny of the five cases of international resource conflict. Through this process, this work was able to identify critical conditions causing nations to engage in these conflicts, as well as crucial consequences to those countries party to the conflicts. Finally, this book has presented a variety of policy recommendations for minimizing resource conflict. Further exploration of this topic appears to be an urgent priority, for while humanity has recently managed to escape catastrophic resource wars, time may well be running out.

Notes

INTRODUCTION

1. David G. Haglund, "The New Geopolitics of Minerals: An Inquiry into the Changing International Significance of Strategic Minerals," *Political Geography Quarterly*, 5 (1986): 227–28; Kenneth A. Dahlberg and others, *Environment and the Global Arena* (Durham, N.C.: Duke University Press, 1985), 80–81; Bohdan O. Szuprowicz, *How to Avoid Strategic Minerals Shortages* (New York: John Wiley and Sons, 1981), 50; and Paul R. Ehrlich and Anne H. Ehrlich, *Population Resources Environment* (San Francisco: W. H. Freeman, 1972), 426.

2. Dahlberg and others, *Environment*, 81.

3. Dennis Pirages, "Scarcity and International Politics," *International Studies Quarterly*, 21 (1977): 563–67; and Dennis Pirages, *Global Ecopolitics* (North Scituate, Mass.: Duxbury Press, 1978), 68–69.

4. Philip W. Quigg, *Environment: The Global Issues* (New York: Foreign Policy Association Headline Series #217, 1973).

5. See, for example, Philip Connelly and Robert Perlman, *The Politics of Scarcity: Resource Conflicts in International Relations* (London: Oxford University Press, 1975); J. E. S. Fawcett and Audrey Parry, *Law and International Resource Conflicts* (London: Oxford University Press, 1981); and Arthur H. Westing, ed., *Global Resources and International Conflict* (New York: Oxford University Press, 1986).

CHAPTER 1

1. Donella H. Meadows and others, *The Limits to Growth* (New York: Universe Books, 1972).

2. William Ophuls, *Ecology and the Politics of Scarcity* (San Francisco: W. H. Freeman, 1977), 169; and Lester R. Brown, *The Global Politics of Resource Scarcity* (Washington D.C.: Overseas Development Council Paper #17, 1974).

3. Dennis Pirages, *Global Ecopolitics* (North Scituate, Mass: Duxbury Press, 1978), chap. 1.

4. Stephen D. Krasner, *Defending the National Interest: Raw Material Investment and U.S. Foreign Policy* (Princeton, N.J.: Princeton University Press, 1978), 350.

5. Julian L. Simon, *The Ultimate Resource* (Princeton, N.J.: Princeton University Press, 1981); and Harold J. Barnett and Chandler Morse, *Scarcity and Growth: The Economics of Natural Resource Availability* (Baltimore: Johns Hopkins University Press, 1963), 8.

6. Ibid., 18–19.

7. Robert C. North, "Toward a Framework for the Analysis of Scarcity and Conflict," *International Studies Quarterly*, 21 (1977): 569–91.

8. Brown, *The Global Politics*, 26; and Nazli Choucri and James D. Bennett, "Population, Resources, and Technology: Political Implications of the Environmental Crisis," in David Kay and Eugene Skolnikoff, eds., *World Eco-Crisis* (Madison, Wis.: University of Wisconsin Press, 1972), 21–26.

9. John Kincaid, "Of Time, Body, and Scarcity: Policy Options and Theoretic Considerations," *International Political Science Review*, 4 (1983): 412.

10. Brown, *The Global Politics*, 26; Ophuls, *Ecology*, 127; Robert Mandel, "Transnational Resource Conflict: The Politics of Whaling," *International Studies Quarterly*, 24 (1980): 101; and Helge Hveem, "Minerals as a Factor in Strategic Policy and Action," in Arthur H. Westing, ed., *Global Resources and International Conflict* (New York: Oxford University Press, 1986), 55.

11. Kincaid, "Of Time," 412.

12. Dennis Pirages, "Scarcity and International Politics," *International Studies Quarterly*, 21 (1977): 563–67; and David G. Haglund, "The New Geopolitics of Minerals: An Inquiry into the Changing International Significance of Strategic Minerals," *Political Geography Quarterly*, 5 (1986): 235.

13. Simon, *The Ultimate Resource*, 41, 221–22, 348.

14. Nathan Rosenberg, "Innovative Responses to Material Shortages," *American Economic Review*, 63 (1973): 111–18.

15. Simon, *The Ultimate Resource*, 22–25; and Barnett and Morse, *Scarcity and Growth*, 8.

16. Charles W. Howe, *Natural Resource Economics: Issues, Analysis, and Policy* (New York: John Wiley and Sons, 1979), 11–12; and Kincaid, "Of Time," 412.

17. Simon, *The Ultimate Resource*, 69, 90–91.

18. Glenn H. Snyder and Paul Diesing, *Conflict Among Nations* (Princeton, N.J.: Princeton University Press, 1977), 467–68; Pirages, "Scarcity," 563–567; and Pirages, *Global Ecopolitics*, Chap. 1, 2–24.

19. Quincy Wright, *A Study of War* (Chicago: University of Chicago Press, 1965), Part 1, 1–100.

20. Dan Morgan, *Merchants of Grain* (New York: Penguin Books, 1980), 343.

21. Robert L. Pfaltzgraff, Jr., "Resource Issues and the Atlantic Community," in Walter F. Hahn and Robert L. Pfaltzgraff, Jr., eds., *Atlantic Community in Crisis* (New York: Pergamon Press, 1979), 307.

22. Geoffrey Kemp, "Scarcity and Strategy," *Foreign Affairs*, 56 (January 1978): 396–97.

23. Nazli Choucri and Robert C. North, *Nations in Conflict: National Growth and International Violence* (San Francisco: W. H. Freeman, 1975); North, "Toward a Framework," 569–91; David W. Orr, "Modernization and Conflict: The Second Image Implications of Scarcity," *International Studies Quarterly*, 21 (1977): 593–618; and Richard J. Barnet, *The Lean Years: Politics in the Age of Scarcity* (New York: Simon and Schuster, 1980), Chap. 8.

24. North, "Toward a Framework," 569–91.

25. Bruce Russett, "Security and the Resources Scramble: Will 1984 Be Like 1914?" *International Affairs*, 58 (Winter 1981/82): 42.

26. Ruth W. Arad and Uzi B. Arad, "Scarce Natural Resources and Potential Conflict," in Ruth W. Arad and others, eds., *Sharing Global Resources* (New York: McGraw-Hill, 1979), 62; and Arthur H. Westing, "An Expanded Concept of International Security" in Westing, *Global Resources*, 183.

27. Kemp, "Scarcity and Strategy," 413.

28. Robert L. Heilbroner, *An Inquiry into the Human Prospect* (New York: W. W. Norton, 1965), 135; C. Fred Bergsten, "The Threat from the Third World," *Foreign Policy*, 11 (1973): 102–24; Orr, "Modernization," 593–618; and Arad and Arad, "Scarce Natural Resources," 28–31.

29. Marvin S. Soroos, *Beyond Sovereignty: The Challenge of Global Policy* (Columbia, S.C.: University of South Carolina Press, 1986), 55.

30. Arad and Arad, "Scarce Natural Resources," 31.

31. Mandel, "Transnational Resource Conflict," 103.

32. Garrett Hardin and John Baden, eds., *Managing the Commons* (San Francisco: W. H. Freeman, 1977); Pfaltzgraff, "Resource Issues," 308; Manus I. Midlarsky, "Scarcity and Inequality: Prologue to the Onset of Mass Revolution," *Journal of Conflict Resolution*, 26 (1982): 3–38; and *The Battle for Natural Resources* (Washington, D.C.: Congressional Quarterly, 1983), Chap. 9.

33. Pfaltzgraff, "Resource Issues," 308.

34. Arad and Arad, "Scarce Natural Resources," 104.

35. John S. Dryzek and Susan Hunter, "Environmental Mediation for International Problems," *International Studies Quarterly*, 31 (1987): 88.

36. Robert Mandel, "Roots of the Modern Interstate Border Dispute," *Journal of Conflict Resolution*, 24 (September 1980): 450.

37. Peter Wallensteen, "Scarce Goods as Political Weapons: The Case of Food," *Journal of Peace Research*, 13 (1976): 277.

38. A. Arbatov and I. Amirov, "Raw Materials Problems in Interstate Conflicts," *International Affairs (Moscow)*, 8 (1984): 103.

39. Kemp, "Scarcity and Strategy," 413.

40. Mandel, "Roots," 446–47.

CHAPTER 2

1. John Kincaid, "Of Time, Body, and Scarcity: Policy Options and Theoretic Considerations," *International Political Science Review*, 4 (1983): 414.

2. Donella H. Meadows and others, *The Limits to Growth* (New York: Universe Books, 1972), Chaps. 3–4, 95–160.

3. William Ophuls, *Ecology and the Politics of Scarcity* (San Francisco: W. H. Freeman, 1977), 187.

4. Richard J. Barnet, *The Lean Years: Politics in the Age of Scarcity* (New York: Simon and Schuster, 1980), Chap. 8, 216–235.

5. Ted Robert Gurr, *Why Men Rebel* (Princeton, N.J.: Princeton University Press, 1970), and Ted Robert Gurr, "On the Political Consequences of Scarcity and Economic Decline," *International Studies Quarterly*, 29 (1985): 61.

6. Johan Galtung, "A Structural Theory of Aggression," *Journal of Peace Research*, 2 (1964): 95–119.

7. Kenneth A. Dahlberg and John W. Bennett, eds., *Natural Resources and People* (Boulder, Colo.: Westview Press, 1986), 355.

8. David W. Orr, "Modernization and Conflict: The Second Image Implications of Scarcity," *International Studies Quarterly*, 21 (1977): 593–618.

9. Robert C. North, "Toward a Framework for the Analysis of Scarcity and Conflict," *International Studies Quarterly*, 21 (1977): 569–91.

10. Barnet, *The Lean Years*, 220.

11. Ophuls, *Ecology*, 188.

12. Peter Wallensteen, "Scarce Goods as Political Weapons: The Case of Food," *Journal of Peace Research*, 13 (1976): 278–79.

13. Stephen J. Andriole, "Research Scarcity and Foreign Policy: Implications for Research and Analysis," *World Affairs*, 139 (1976): 17–26; Ophuls, *Ecology*, 214–15; and John S. Dryzek and Susan Hunter, "Envi-

ronmental Mediation for International Problems," *International Studies Quarterly*, 31 (1987): 87.

14. Orr, "Modernization," 593–618.

15. Lester R. Brown, "Redefining National Security," in Lester R. Brown and others, eds., *State of the World 1986* (New York: W. W. Norton, 1986), 195–97.

16. Geoffrey Kemp, "Scarcity and Strategy," *Foreign Affairs*, 56 (January 1978): 407.

17. Ibid., 409–12.

18. Harold J. Barnett and Chandler Morse, *Scarcity and Growth: The Economics of Natural Resource Availability* (Baltimore: Johns Hopkins University Press, 1963), 10.

19. Kemp, "Scarcity and Strategy," 398.

20. Ibid., 399; and Barnett and Morse, *Scarcity and Growth*, 4, 7.

21. Barnett and Morse, *Scarcity and Growth*, 5.

22. Helge Hveem, "Minerals as a Factor in Strategic Policy and Action," in Arthur H. Westing, ed., *Global Resources and International Conflict* (New York: Oxford University Press, 1986), 56.

23. Kincaid, "Of Time," 402.

24. Richard A. Falk, *This Endangered Planet* (New York: Random House, 1972), Chap. 3, 37–92; and Gurr, "On the Political Consequences," 54.

25. Mancur Olson, "Rapid Growth as a Destabilizing Force," *Journal of Economic History*, 23 (1963): 529–52; Quincy Wright, *A Study of War* (Chicago: University of Chicago Press, 1965), Part 1, 1–100; and Robert Gilpin, *War and Change in World Politics* (New York: Cambridge University Press, 1981).

26. Barnet, *The Lean Years*, Chap. 6, 151–190.

27. Kenneth A. Dahlberg and others, *Environment and the Global Arena* (Durham, N.C.: Duke University Press, 1985), 33–35.

28. Arthur A. Stein and Bruce M. Russett, "Evaluating War: Outcomes and Consequences," in Ted Robert Gurr, ed., *Handbook of Political Conflict* (New York: Free Press, 1980), 409.

29. Johan Galtung, *Environment, Development and Military Activity* (Oslo: Universitetsforlaget, 1982), 34.

30. Gurr, "On the Political Consequences," 71.

31. Barnett and Morse, *Scarcity and Growth*, 12.

32. Ibid., 8.

33. Paul R. Ehrlich and Anne H. Ehrlich, *The End of Affluence* (New York: Ballantine, 1974), 108–13.

34. Ruth W. Arad and Uzi B. Arad, "Scarce Natural Resources and Potential Conflict," in Ruth W. Arad and others, eds., *Sharing Global Resources* (New York: McGraw-Hill, 1979), 28.

35. Robert L. Pfaltzgraff, Jr., "Resource Issues and the Atlantic Community," in Walter F. Hahn and Robert L. Pfaltzgraff, Jr., eds., *Atlantic Community in Crisis* (New York: Pergamon Press, 1979), 298.

36. Bohdan O. Szuprowicz, *How to Avoid Strategic Minerals Shortages* (New York: John Wiley and Sons, 1981), ix, 55.

37. Arad and Arad, "Scarce Natural Resources," 87–96.

38. Edward J. Woodhouse, "Re-Visioning the Future of the Third World: An Ecological Perspective on Development," *World Politics*, 25 (1972): 1–33.

39. Ibid.

40. Brown, "Redefining National Security," 204.

41. Szuprowicz, *How to Avoid*, 49.

42. Stein and Russett, *Evaluating War*, 411–13.

43. Galtung, *Environment*, 43.

44. Ophuls, *Ecology*, 145; and Gurr, "On the Political Consequences," 58–59.

45. Manus I. Midlarsky, "Scarcity and Inequality: Prologue to the Onset of Mass Revolution," *Journal of Conflict Resolution*, 26 (1982): 34.

46. Ophuls, *Ecology*, Chap. 6, 184–199.

47. Ibid., 145; Marvin S. Soroos, *Beyond Sovereignty: The Challenge of Global Policy* (Columbia, S.C.: University of South Carolina Press, 1986), 54; and Arthur H. Westing, "An Expanded Concept of International Security," in Westing, *Global Resources*, 195.

48. Kincaid, "Of Time," 412.

49. Ehrlich and Ehrlich, *The End*, 94.

50. Szuprowicz, *How to Avoid*, 33.

51. Midlarsky, "Scarcity," 33.

52. Ophuls, *Ecology*, Chap. 8, 222–248.

53. Stein and Russett, *Evaluating War*, 414–418.

54. Galtung, *Environment*, 43.

55. Arthur A. Stein, *The Nation at War* (Baltimore: Johns Hopkins University Press, 1980), 92.

56. Philip W. Quigg, *Environment: The Global Issues* (New York: Foreign Policy Association Headline Series #217, 1973).

57. Edward Friedland, Paul Seabury, and Aaron Wildavsky, *The Great Detente Disaster: Oil and the Decline of American Foreign Policy* (New York: Basic Books, 1975), 100.

58. Ehrlich and Ehrlich, *The End*, 93–95; Harold Sprout and Margaret Sprout, *Multiple Vulnerabilities: The Context of Environmental Repair and Protection* (Princeton, N.J.: Center of International Studies monograph, #40, 1974); and Orr, "Modernization," 593–618.

59. Robert O. Keohane and Joseph S. Nye, *Power and Interdependence: World Politics in Transition* (Boston: Little, Brown, 1977), 15.

60. Robert Wilson, "Natural Resources: Dependency and Vulnerability," in Uri Ra'anan and Charles M. Perry, eds., *Strategic Minerals and International Security* (Washington, D.C.: Pergamon, 1985), 25.

61. Barry M. Blechman, *National Security and Strategic Minerals* (Boulder, Colo.: Westview Press, 1985), 2–3.

62. Pfaltzgraff, "Resource Issues," 300.

63. Robert Legvold, "The Strategic Implications of the Soviet Union's Non-Fuel Mineral Resources Policy," *Journal of Resource Management and Technology*, 12 (1983): 55; and David G. Haglund, "Strategic Minerals: A Conceptual Analysis," *Resources Policy*, 10 (1984): 149–50.

64. Hanns Maull, "Oil and Influence: The Oil Weapon Examined," in Klaus Knorr and Frank N. Trager, eds., *Economic Issues and National Security* (Lawrence, Kan.: Regents Press, 1977), 280.

65. Sprout and Sprout, *Multiple Vulnerabilities*; Dennis Pirages, *Global Ecopolitics* (North Scituate, Mass.: Duxbury Press, 1978), 189–200.

66. Ehrlich and Ehrlich, *The End*, 94, 107.

67. Barnet, *The Lean Years*, Chap. 11, 295–317.

68. Julian L. Simon, *The Ultimate Resource* (Princeton, N.J.: Princeton University Press, 1981), 270–71.

69. Stein and Russett, *Evaluating War*, 412.

70. Galtung, *Environment*, 43.

71. Glenn H. Snyder and Paul Diesing, *Conflict Among Nations* (Princeton, N.J.: Princeton University Press, 1977), 170, 429.

72. Ophuls, *Ecology*, 134; Kincaid, "Of Time," 401.

73. Dennis Meadows and Jorgen Randers, "Adding the Time Dimension in Environmental Policy," in David Kay and Eugene Skolnikoff, eds., *World Eco-Crisis* (Madison, Wis.: University of Wisconsin Press, 1972), 47–66.

74. Leonard Berry and Douglas L. Johnson, "Geographical Approaches to Environmental Change: Assessing Human Impacts on Global Resources," in Dahlberg and Bennett, eds., *Natural Resources*, 80.

75. Meadows and Randers, "Adding the Time Dimension," 47–66.

76. Maull, "Oil," 281.

CHAPTER 3

1. Alexander L. George and Richard Smoke, *Deterrence in American Foreign Policy: Theory and Practice* (New York: Columbia University Press, 1974), 95–97.

2. Dennis Pirages, *Global Ecopolitics* (North Scituate, Mass.: Duxbury Press, 1978).

CHAPTER 4

1. Philip W. Quigg, *Environment: The Global Issues* (New York: Foreign Policy Association Headline Series #217, 1973); and Garrett Hardin and John Baden, eds., *Managing the Commons* (San Francisco: W. H. Freeman, 1977).

2. Lester R. Brown, *The Global Politics of Resource Scarcity* (Washington, D.C.: Overseas Development Council Paper #17, 1974), 26–28.

3. Daniel Fife, "Killing the Goose," in Hardin and Baden, eds., *Managing the Commons*, 76–81.

4. Marc Leepson, "Whaling: End of an Era," *Editorial Research Reports*, 2 (September 27, 1985): 725.

5. Robert Mandel, "Transnational Resource Conflict: The Politics of Whaling," *International Studies Quarterly*, 24 (1980): 105.

6. Leepson, "Whaling," 730.

7. James P. Sterba, "Whale Doomed, Ecologists Say, But Industry Sees Fear as Myth," *New York Times* (November 30, 1971): 23.

8. Leepson, "Whaling," 720.

9. Ibid.

10. "Soviets End Whaling," *Sunday Oregonian* (May 24, 1987): A4.

11. Leepson, "Whaling," 722.

12. Eugene Moosa, "Japanese Whalers Feel 'Singled Out'," *Sunday Oregonian* (April 28, 1985): A10.

13. Leepson, "Whatling," 724–25.

14. Moosa, "Japanese Whalers," A10; and John Burgess, "Japan to Renew Whale Hunt as Research Effort," *Washington Post* (April 8, 1987): A1.

15. Thomas W. Netter, "Conservationists Fear Breakdown in Whaling Moratorium," *New York Times* (August 19, 1986): C3.

16. Burgess, "Japan," A1.

17. Mandel, "Transnational Resource Conflict," 109.

18. David O. Hill, "Vanishing Giants," *Audubon*, 77 (1975): 86.

19. Ann Cottrell Free, "Will They A-Whaling Go?" *Washington Star-News* (June 23, 1974): C–2.

20. Fox Butterfield, "Appeal to Tokyo Asks Whaling Ban," *New York Times* (June 6, 1974): 7.

21. Mandel, "Transnational Resource Conflict," 112.

22. Friends of the Earth, *The Whale Manual* (San Francisco: Friends of the Earth Books, 1978), iv.

23. Interview with Liz Tilbury, Greenpeace Oregon, August 3, 1978.

24. Ibid.; and Mandel, "Transnational Resource Conflict," 109–11.

25. B. J. Williams, "Kamikaze Conservationists: Radicals Take to the Sea," *Northwest Magazine*, 22 (August 16, 1987): 6–11; and David Day, *The Whale War* (San Francisco: Sierra Club Books, 1987).

26. Faith McNulty, "Profiles: Lord of the Fish," *New Yorker*, 49 (August 6, 1973): 40.

27. Sterba, "Whale Doomed," 23.

28. Leepson, "Whaling," 722.

29. Rempei Komatsu, *Whaling and Japan* (Tokyo: Japan Institute of International Affairs, 1974), 11; and Andrew H. Malcolm, "Japanese Whaling Ports Are Distressed as Quotas Cut and Industry Slumps," *New York Times* (June 14, 1976): 10.

30. Moosa, "Japanese Whalers," A10; and Burgess, "Japan," A43.

31. Moosa, "Japanese Whalers," A10.

32. Irston R. Barnes, "To Save a Whale," *Washington Post* (March 17, 1974): E14; Hill, "Vanishing Giants," 87; and Center for Action on Endangered Species, *The Whalebook* (Washington, D.C.: Center for Environmental Education, 1978), 27–30.

33. "Soviet Announces It Plans to Reduce Antarctic Whaling," *New York Times* (June 24, 1975): 7.

34. Komatsu, *Whaling*, 25.

35. Malcolm, "Japanese Whaling," 10; and Leepson, "Whaling," 723.

36. *Are Whales Really Threatened with Extinction?* (Tokyo: Japan Whaling Association, 1973), 22–23; Mikhail Ivashin, "Sensible Whaling," unpublished paper, December 13, 1977, Soviet Embassy, Washington, D.C.; and Moosa, "Japanese Whalers," A10.

37. Center for Action on Endangered Species, *The Whalebook*, vi, 3; and Friends of the Earth, *The Whale Manual*, 16–17, 46.

38. Leepson, "Whaling," 734.

39. Komatsu, *Whaling*, 28; and Ivan Nikonorov and Mikhail Ivashin, "The Soviet Position on Whaling," unpublished paper, August 17, 1977, Soviet Embassy, Washington, D.C. 4–5.

40. Ray Gambell, "Why All the Fuss About Whales" *New Scientist*, 54 (June 22, 1972): 674–76; and Leepson, "Whaling," 735.

41. *The Whaling Controversy: Japan's Position and Proposals* (Tokyo: Japan Whaling Association, 1978), 2; and John R. Schmidhauser and George O. Totten III, eds., *The Whaling Issue in U.S.-Japan Relations* (Boulder, Colo.: Westview Press, 1978), 5.

42. Nikonorov and Ivashin, "The Soviet Position," 4–5.

43. Tom Garrett, "A 'Final Solution' for the Whales," *New York Times* (August 15, 1971), 29; and Schmidhauser and Totten, *The Whaling Issue*, 5.

44. Joanna Gordon Clark, Angela King, and John Burton, "Whales: Time for a Fresh Start," *New Scientist*, 65 (January 23, 1975): 206, 208.

45. Hill, *Vanishing Giants*, 90.

46. *Scientific Whaling Conservation: An International Responsibility* (Tokyo: Japan Echo, 1975), 1.

47. Nikonorov and Ivashin, "The Soviet Position," 1–2.
48. "Japan's Whale Agony," *Washington Star-News* (July 8, 1974): A–10; Schmidhauser and Totten, *The Whaling Issue*, 4; and Moosa, "Japanese Whalers," A10.
49. Leepson, "Whaling," 720.
50. Mandel, "Transnational Resource Conflict," 116–17.
51. Leepson, "Whaling," 735.
52. "Soviets End Whaling," A4.
53. Moosa, "Japanese Whalers," A10; and Leepson, "Whaling," 723.
54. Mandel, "Transnational Resource Conflict," 111; and Williams, "Kamikaze Conservationists," 6–11.
55. Interview with Tilbury; and Mandel, "Transnational Resource Conflict," 112.

CHAPTER 5

1. Richard J. Barnet, *The Lean Years: Politics in the Age of Scarcity* (New York: Simon and Schuster, 1980), 66, 73.
2. Christopher Flavin, "Moving Beyond Oil," in Lester R. Brown and others, eds., *State of the World 1986* (New York: W. W. Norton, 1986), 87.
3. Hanns Maull, "Oil and Influence: The Oil Weapon Examined," in Klaus Knorr and Frank N. Trager, eds., *Economic Issues and National Security* (Lawrence, Kan.: Regents Press, 1977), 262.
4. Anthony Sampson, *The Seven Sisters: The Great Oil Companies and the World They Shaped* (New York: Bantam Books, 1976), 287, 290.
5. Ibid., 279.
6. Ibid., 330.
7. Roy E. Licklider, "The Failure of the Arab Oil Weapon in 1973–1974," *Comparative Strategy*, (1982): 374.
8. Benjamin Shwadran, *Middle East Oil Crises Since 1973* (Boulder, Colo.: Westview Press, 1986), 65.
9. Sampson, *The Seven Sisters*, 301.
10. Jordan J. Paust and Albert P. Blaustein, *The Arab Oil Weapon* (Dobbs Ferry, N.Y.: Oceana Publications, 1977), 84.
11. Raymond Vernon, "An Interpretation," *Daedalus*, 104 (1975): 2–3.
12. Maull, "Oil," 263.
13. Peter R. Odell, *Oil and World Politics*, 5th ed. (New York: Penguin Books, 1979), 220.
14. Klaus Knorr, "The Limits of Economic and Military Power," *Daedalus*, 104 (1975): 233.
15. Ian Smart, "Uniqueness and Generality," *Daedalus*, 104 (1975): 270.
16. Sampson, *The Seven Sisters*, 309.

17. Ibid., 311–12.
18. Knorr, "The Limits," 230–31.
19. Licklider, "The Failure," 366–67.
20. Shwadran, *Middle East Oil*, 59.
21. Odell, *Oil*, 224–25.
22. Shwadran, *Middle East Oil*, 58.
23. Edith Penrose, "The Development of Crisis," *Daedalus*, 104 (1975): 52.
24. Odell, *Oil*, 239–40; and Flavin, "Moving Beyond Oil," 87.
25. Odell, *Oil*, 225; and Shwadran, *Middle East Oil*, 81.
26. Gerald A. Pollack, "The Economic Consequences of the Energy Crisis," *Foreign Affairs*, 52 (1974): 461.
27. Shwadran, *Middle East Oil*, 57.
28. Maull, "Oil," 265.
29. Ibid., 267.
30. Zuhayr Mikdashi, "The OPEC Process," *Daedalus*, 104 (1975): 211–12.
31. Licklider, "The Failure," 373.
32. Smart, "Uniqueness," 263.
33. Sampson, *The Seven Sisters*, 329.
34. Maull, "Oil," 259.
35. Sampson, *The Seven Sisters*, 317.
36. Ibid., 318.
37. Shwadran, *Middle East Oil*, 65.
38. Ibid., 57.
39. Knorr, "The Limits," 242.
40. Raymond Vernon, "Distribution of Power," *Daedalus*, 104 (1975): 254.
41. Sampson, *The Seven Sisters*, 320–21.
42. Smart, "Uniqueness," 274.
43. Maull, "Oil," 268–69.
44. Licklider, "The Failure," 370–71.
45. Mikdashi, "The OPEC Process," 212.
46. Shwadran, *Middle East Oil*, 92–93, 111.
47. Maull, "Oil," 270–71.
48. Shwadran, *Middle East Oil*, 48–63.

CHAPTER 6

1. Peter Wallensteen, "Scarce Goods as Political Weapons: The Case of Food," *Journal of Peace Research*, 13 (1976): 279.
2. Raymond F. Hopkins and Donald J. Puchala, *Global Food Interde-*

pendence: Challenge to American Foreign Policy (New York: Columbia University Press, 1980), 1.

3. Lester R. Brown, "Sustaining World Agriculture," in Lester R. Brown and others, eds., *State of the World 1987* (New York: W. W. Norton, 1987), 133.

4. Ibid.

5. Hopkins and Puchala, *Global Food*, 3.

6. Richard J. Barnet, *The Lean Years: Politics in the Age of Scarcity* (New York: Simon and Schuster, 1980), Chap. 6, 151–190.

7. Dan Morgan, *Merchants of Grain* (New York: Penguin Books, 1980), 13.

8. Barnet, *The Lean Years*, 152.

9. Trudy Huskamp Peterson, "Sales Surpluses, and the Soviets: A Study in Political Economy," *Policy Studies Journal*, 6 (1978): 531.

10. Morgan, *Merchants*, 208.

11. Nick Butler, "The US Grain Weapon: Could It Boomerang?" *The World Today*, 39 (1983): 52.

12. Morgan, *Merchants*, 200.

13. Ibid., 336.

14. Ibid., 334.

15. Robert L. Paarlberg, "Lessons of the Grain Embargo," *Foreign Affairs*, 59 (1980): 146; and Morgan, *Merchants*, 352, 359–60.

16. Barnet, *The Lean Years*, 157.

17. Morgan, *Merchants*, 357.

18. Roger B. Porter, "The U.S.-U.S.S.R. Grain Agreement: Some Lessons for Policymakers," *Public Policy*, 29 (1981): 549.

19. Morgan, *Merchants*, 360.

20. Paarlberg, "Lessons," 146.

21. Ibid., 144–45.

22. Richard Gilmore, "Grain in the Bank," *Foreign Policy*, 38 (1980): 168.

23. Paarlberg, "Lessons," 144.

24. Ibid., 147.

25. Dale E. Hathaway, "The Internationalization of U.S. Agriculture," in Emery N. Castle and Kent A. Price, eds., *U.S. Interests and Global Natural Resources: Energy, Minerals, Food* (Washington, D.C.: Resources for the Future, 1983), 91.

26. Paarlberg, "Lessons," 151.

27. Hathaway, "The Internationalization," 92.

28. Morgan, *Merchants*, 344.

29. Cheryl Christensen, "Food and National Security," in Klaus Knorr and Frank N. Trager, eds., *Economic Issues and National Security* (Lawrence,

Kan.: Regents Press, 1977), 293; Robert L. Paarlberg, "Food, Oil, and Coercive Resource Power," *International Security*, 3 (1978): 3; and Hopkins and Puchala, *Global Food*, 151.

30. Butler, "The US Grain Weapon," 52.

31. Wallensteen, "Scarce Goods," 287–94.

32. Paarlberg, "Lessons," 160.

33. Morgan, *Merchants*, 346.

34. Ibid., 347.

35. Hopkins and Puchala, *Global Food*, xv.

36. Christensen, "Food," 299–300; Barnet, *The Lean Years*, 158–59; and Morgan, *Merchants*, 359.

37. Wallensteen, "Scarce Goods," 278–79.

38. Paarlberg, "Lessons," 145.

39. Hopkins and Puchala, *Global Food*, 152, 171.

40. Butler, "The US Grain Weapon," 52–53.

41. Christensen, "Food," 299–302.

42. Gilmore, "Grain," 171–72.

43. Hopkins and Puchala, *Global Food*, xvi.

44. Paarlberg, "Lessons," 149.

45. Paarlberg, "Food," 6.

46. Morgan, *Merchants*, 363.

47. Hathaway, "The Internationalization," 91.

48. Roy D. Laird, "Grain as a Foreign Policy Tool in Dealing with the Soviets: A Contingency Plan," *Policy Studies Journal*, 6 (1978): 534.

49. Butler, "The US Grain Weapon," 55.

50. Christensen, "Food," 302; Morgan, *Merchants*, 364; and Hathaway, "The Internationalization," 91.

51. Laird, "Grain," 535.

52. Butler, "The US Grain Weapon," 55.

53. Ibid.; and Hathaway, "The Internationalization," 92–95.

54. Barbara Insel, "A World Awash in Grain," *Foreign Affairs*, 63 (1985): 903.

CHAPTER 7

1. James T. Bennett and Walter E. Williams, *Strategic Minerals: The Economic Impact of Supply Disruptions* (Washington, D.C.: Heritage Foundation, 1981), v.

2. Ibid.

3. Bension Varon and Kenji Takeuchi, "Developing Countries and Non-Fuel Minerals," *Foreign Affairs*, 52 (1974): 497.

4. Ibid., 497–98.

5. Richard J. Barnet, *The Lean Years: Politics in the Age of Scarcity* (New York: Simon and Schuster, 1980), 114; and David G. Haglund, "The New Geopolitics of Minerals: An Inquiry into the Changing International Significance of Strategic Minerals," *Political Geography Quarterly*, 5 (1986): 233.

6. Haglund, "The New Geopolitics," 225.

7. Bennett and Williams, *Strategic Minerals*, v.

8. Bohdan O. Szuprowicz, *How to Avoid Strategic Minerals Shortages* (New York: John Wiley and Sons, 1981), 90.

9. Michael Shafer, "Mineral Myths," *Foreign Policy*, 47 (1982): 163.

10. Haglund, "The New Geopolitics," 221.

11. William Schneider, Jr., "Strategic Minerals: International Considerations," in Uri Ra'anan and Charles M. Perry, eds., *Strategic Minerals and International Security* (Washington, D.C.: Pergamon, 1985), 70.

12. Rae Weston, *Strategic Minerals: A World Survey* (Totowa, N.J.: Rowman and Allanheld, 1984), 111.

13. Schneider, "Strategic Minerals," 69.

14. Barry M. Blechman, *National Security and Strategic Minerals* (Boulder, Colo.: Westview Press, 1985), xiii.

15. Shafer, "Mineral Myths," 161.

16. Weston, *Strategic Minerals*, 107.

17. Barnet, *The Lean Years*, 122–23.

18. Blechman, *National Security*, ix.

19. Haglund, "The New Geopolitics," 236.

20. Ibid.

21. Barnet, *The Lean Years*, 125.

22. Alwyn H. King, "The Strategic Minerals Problem: Our Domestic Options," *Parameters*, 12 (1982): 47; and Weston, *Strategic Minerals*, 2–4.

23. W. C. J. Van Rensburg, *Strategic Minerals: Major Mineral-Exporting Regions of the World*, vol. 1 (Englewood Cliffs, N.J.: Prentice-Hall, 1986), 2.

24. Szuprowicz, *How to Avoid*, 136–37.

25. Schneider, "Strategic Minerals," 69–70.

26. Shafer, "Mineral Myths," 159–60.

27. Oye Ogunbadejo, *The International Politics of Africa's Strategic Minerals* (Westport, Conn.: Greenwood Press, 1985), 195–98.

28. Robert Legvold, "The Strategic Implications of the Soviet Union's Non-Fuel Minerals Policy," *Journal of Resource Management and Technology*, 12 (1983): 50, 53–54.

29. Szuprowicz, *How to Avoid*, 50; and Haglund, "The New Geopolitics," 229.

30. King, "The Strategic Minerals Problem," 44.

31. Haglund, "The New Geopolitics," 225.

32. Ibid., 227.

33. Van Rensburg, *Strategic Minerals*, vol. 1, xiii.

34. Ogunbadejo, *The International Politics*, 2–3.

35. For example, Weston, *Strategic Minerals*, 92–93.

36. Bennett and Williams, *Strategic Minerals*, 56.

37. R. Daniel McMichael, "Strategic Materials: The Public Policy Process," in Ra'anan and Perry, eds., *Strategic Minerals*, 8.

38. Shafer, "Mineral Myths," 155–59, 162–63.

39. Legvold, "The Strategic Implications," 48; Weston, *Strategic Minerals*, 151; and W. C. J. Van Rensburg, *Strategic Minerals: Major Mineral-Consuming Regions of the World*, vol. 2 (Englewood Cliffs, N.J.: Prentice-Hall, 1986), x.

40. Weston, *Strategic Minerals*, 141.

41. Bennett and Williams, *Strategic Minerals*, 41.

42. Weston, *Strategic Minerals*, 108–109.

43. Ogunbadejo, *The International Politics*, 2–8.

44. Van Rensburg, *Strategic Minerals*, vol. 1, 6.

45. Ibid.

46. Blechman, *National Security*, 7.

47. McMichael, "Strategic Materials," 5.

48. Van Rensburg, *Strategic Minerals*, vol. 1, 6.

CHAPTER 8

1. Kenneth A. Dahlberg and others, *Environment and the Global Arena* (Durham, N.C.: Duke University Press, 1985), 70–71.

2. Marvin S. Soroos, *Beyond Sovereignty: The Challenge of Global Policy* (Columbia, S.C.: University of South Carolina Press, 1986), 57.

3. William Ophuls, *Ecology and the Politics of Scarcity* (San Francisco: W. H. Freeman, 1977), 74–76.

4. For example, David Schwartz, "On the Ecology of Political Violence: 'The Long Hot Summer' as a Hypothesis," *American Behavioral Scientist*, 11 (1968): 24–28.

5. Philip W. Quigg, *Environment: The Global Issues* (New York: Foreign Policy Association Headline Series #217, 1973), 4–7; and John S. Dryzek and Susan Hunter, "Environmental Mediation for International Problems," *International Studies Quarterly*, 31 (1987), 87.

6. Johan Galtung, *Environment, Development and Military Activity* (Oslo: Universitetsforlaget, 1982), 34, 63.

7. Richard J. Barnet, *The Lean Years: Politics in the Age of Scarcity* (New York: Simon and Schuster, 1980), 85.

8. Bennett Ramberg, "Lessons from Chernobyl," *Foreign Affairs*, 65 (1986/87): 320–21.

9. Christopher Flavin, "Reassessing Nuclear Power," in Lester R. Brown and others, eds., *State of the World 1987* (New York: W. W. Norton, 1987), 68.

10. Ramberg, "Lessons," 305.

11. Ibid., 307–9.

12. Jon Van, "Disaster Fuels Doubt over Nuclear Power," *Sunday Oregonian* (April 26, 1987): A10.

13. Ramberg, "Lessons," 310, 317.

14. Sharon Begley and others, "The Lessons of Chernobyl," *Newsweek*, 109 (April 27, 1987): 56; and Flavin, "Reassessing Nuclear Power," 57.

15. Flavin, "Reassessing Nuclear Power," 59.

16. Begley and others, "The Lessons," 56.

17. Ibid., 56–57.

18. Flavin, "Reassessing Nuclear Power," 57.

19. Ibid., 63.

20. Erik P. Hoffmann, "Nuclear Deception: Soviet Information Policy," *Bulletin of the Atomic Scientists*, 42 (1986): 32.

21. Flavin, "Reassessing Nuclear Power," 62.

22. Hoffmann, "Nuclear Deception," 32; and Flavin, "Reassessing Nuclear Power," 62–63.

23. Hoffmann, "Nuclear Deception," 33.

24. Ibid., 36.

25. Ramberg, "Lessons," 317.

26. Hoffmann, "Nuclear Deception," 36.

27. Flavin, "Reassessing Nuclear Power," 66.

28. Ibid., 66–67.

29. John Greenwald, "More Fallout from Chernobyl," *Time*, 127 (May 19, 1986): 44.

30. Kevin Costelloe, "Europe under Cloud of Doubt on Anniversary of Accident," *Sunday Oregonian* (April 26, 1987): A11.

31. Flavin, "Reassessing Nuclear Power," 57, 61.

32. Ibid., 61.

33. Costelloe, "Europe," A11.

34. Ramberg, "Lessons," 317–18.

35. Flavin, "Reassessing Nuclear Power," 63.

36. Van, "Disaster," A1, A10.

37. Greenwald, "More Fallout," 46.

38. Hoffmann, "Nuclear Deception," 36.

39. Flavin, "Reassessing Nuclear Power," 63–65.

40. Costelloe, "Europe," A11.

41. Ramberg, "Lessons," 315.
42. Flavin, "Reassessing Nuclear Power," 65.
43. Ibid., 75.
44. Alvin M. Weinberg, "A Nuclear Power Advocate Reflects on Chernobyl," *Bulletin of the Atomic Scientists*, 42 (1986): 57.
45. David A. V. Fischer, "The International Response," *Bulletin of the Atomic Scientists*, 42 (1986): 47–48.
46. Flavin, "Reassessing Nuclear Power," 58, 61–62.
47. Fischer, "The International Response," 46–47.
48. Ibid., 46.
49. Ibid.
50. Ibid., 47.
51. Begley and others, "The Lessons," 59.

CHAPTER 10

1. Arthur H. Westing, "An Expanded Concept of International Security," in Arthur H. Westing, ed., *Global Resources and International Conflict* (New York: Oxford University Press, 1986), 195–96.

2. John R. Schmidhauser and George O. Totten III, eds., *The Whaling Issue in U.S.-Japan Relations* (Boulder, Colo.: Westview Press, 1978), 265–73; Robert Mandel, "Transnational Resource Conflict: The Politics of Whaling," *International Studies Quarterly*, 24 (1980): 115; and Marc Leepson, "Whaling: End of an Era," *Editorial Research Reports*, 2 (September 27, 1985): 734–35.

3. Joanna Clark, Angela King, and John Burton, "Whales: Time for a Fresh Start," *New Scientist*, 65 (January 23, 1975): 208; Schmidhauser and Totten, *The Whaling Issue*, 266; John S. Dryzek and Susan Hunter, "Environmental Mediation for International Problems," *International Studies Quarterly*, 31 (1987): 99; and Mandel, "Transnational Resource Conflict," 121.

4. Joel Darmstadter and Hans H. Landsberg, "The Economic Background," *Daedalus*, 104 (1975): 35–36; Ulf Lantzke, "The OECD and its International Energy Agency," *Daedalus*, 104 (1975): 219; Jordan J. Paust and Albert P. Blaustein, *The Arab Oil Weapon* (Dobbs Ferry, N.Y.: Oceana Publications, 1977), 15, 20; and Anthony Sampson, *The Seven Sisters: The Great Oil Companies and the World They Shaped* (New York: Bantam Books, 1976), 329.

5. Darmstadter and Landsberg, "The Economic Background," 36; Lantzke, "The OECD," 220, 224–27; Raymond Vernon, "An Interpretation," *Daedalus*, 104 (1975): 5; Peter R. Odell, *Oil and World Politics*, 5th ed. (New York: Penguin Books, 1979), 242, 246; Richard J. Barnet, *The*

Lean Years: Politics in the Age of Scarcity (New York: Simon and Schuster, 1980), 107; and Christopher Flavin, "Moving Beyond Oil," in Lester R. Brown and others, eds., *State of the World 1986* (New York: W. W. Norton, 1986), 84.

6. Nick Butler, "The US Grain Weapon: Could It Boomerang?" *The World Today*, 39 (1983): 58–59.

7. Raymond F. Hopkins and Donald J. Puchala, *Global Food Interdependence: Challenge to American Foreign Policy* (New York: Columbia University Press, 1980), 153–87; Richard Gilmore, "Grain in the Bank," *Foreign Policy*, 38 (1980): 177–81; and Peter Wallensteen, "Scarce Goods as Political Weapons: The Case of Food," *Journal of Peace Research*, 13 (1976): 295.

8. Alwyn H. King, "The Strategic Minerals Problem: Our Domestic Options," *Parameters*, 12 (1982): 47–50; Michael Shafer, "Mineral Myths," *Foreign Policy*, 47 (1982): 163–71; Barry M. Blechman, *National Security and Strategic Minerals* (Boulder, Colo.: Westview Press, 1985), xv, 36–37; and Uri Ra'anan and Charles M. Perry, eds., *Strategic Minerals and International Security* (Washington, D.C.: Pergamon, 1985), vii.

9. Amos A. Jordon, Robert A. Kilmarx, and Dan Haendel, "The U.S. Strategic Minerals Stockpile: Remedy for Increasing Vulnerability?" *Comparative Strategy*, 1 (1979): 325–28; King, "The Strategic Minerals Problem," 49–50; and Ra'anan and Perry, *Strategic Minerals*, vii.

10. David A. V. Fischer, "The International Response," *Bulletin of the Atomic Scientists*, 42 (1986): 48; and Bennett Ramberg, "Lessons from Chernobyl," *Foreign Affairs*, 65 (1986/87): 326–28.

11. Alvin M. Weinberg, "A Nuclear Power Advocate Reflects on Chernobyl," *Bulletin of the Atomic Scientists*, 42 (1986): 59–60; and Christopher Flavin, "Reassessing Nuclear Power," in Lester R. Brown and others, eds., *State of the World 1987* (New York: W. W. Norton, 1987), 58.

12. One specific resource focus suggested for intensified American government-corporation consultation concerns predicting political risks in overseas extractive industries. See Robert Mandel, "Predicting Overseas Political Instability: Perspectives of the Government Intelligence and Multinational Business Communities," *Conflict Quarterly*, 7 (Spring 1988), forthcoming.

13. Dryzek and Hunter, "Environmental Mediation," 88–95.

14. William Ophuls, *Ecology and the Politics of Scarcity* (San Francisco: W. H. Freeman, 1977), chap. 5.

15. Garrett Hardin and John Baden, eds., *Managing the Commons* (San Francisco: W. H. Freeman, 1977), chap. 13.

16. Julian L. Simon, *The Ultimate Resource* (Princeton University Press, 1981), 149–51.

Selected Bibliography

Andriole, Stephen J. "Resource Scarcity and Foreign Policy: Implications for Research and Analysis." *World Affairs*, 139 (1976): 17–26.

Arad, Ruth and others, eds. *Sharing Global Resources*. New York: McGraw-Hill, 1979.

Barnet, Richard J. *The Lean Years: Politics in the Age of Scarcity*. New York: Simon and Schuster, 1980.

Brown, Lester R. and others, eds. *State of the World 1987*. New York: W. W. Norton, 1987.

Castle, Emery N. and Kent A. Price, eds. *U.S. Interests and Global Natural Resources: Energy, Minerals, Food*. Washington, D.C.: Resources for the Future, 1983.

Choucri, Nazli and Robert C. North. *Nations in Conflict: National Growth and International Violence*. San Francisco: W. H. Freeman, 1975.

Connelly, Philip and Robert Perlman. *The Politics of Scarcity: Resource Conflicts in International Relations*. London: Oxford University Press, 1975.

Dahlberg, Kenneth A. and others. *Environment and the Global Arena*. Durham, N.C.: Duke University Press, 1985.

Dryzek, John S. and Susan Hunter. "Environmental Mediation for International Problems." *International Studies Quarterly*, 31 (March 1987): 87–102.

Galtung, Johan. *Environment, Development and Military Activity*. Oslo: Universitetsforlaget, 1982.

Gurr, Ted Robert. "On the Political Consequences of Scarcity and Economic Decline." *International Studies Quarterly*, 29 (March 1985): 51–75.

Hardin, Garrett and John Baden, eds. *Managing the Commons*. San Francisco: W. H. Freeman, 1977.

Kemp, Geoffrey. "Scarcity and Strategy." *Foreign Affairs*, 56 (January 1978): 396–414.

Kincaid, John. "Of Time, Body, and Scarcity: Policy Options and Theoretic Considerations." *International Political Science Review*, 4 (1983): 401–16.

Knorr, Klaus and Frank N. Trager, eds. *Economic Issues and National Security*. Lawrence, Kan.: Regents Press, 1977.

Mandel, Robert. "Transnational Resource Conflict: The Politics of Whaling." *International Studies Quarterly*, 24 (March 1980): 99–127.

North, Robert C. "Toward a Framework for the Analysis of Scarcity and Conflict." *International Studies Quarterly*, 21 (December 1977): 569–91.

Ophuls, William. *Ecology and the Politics of Scarcity*. San Francisco: W. H. Freeman, 1977.

Paarlberg, Robert L. "Food, Oil, and Coercive Resource Power." *International Security*, 3 (Fall 1978): 3–19.

Pirages, Dennis. *Global Ecopolitics*. North Scituate, Mass.: Duxbury Press, 1978.

Simon, Julian L. *The Ultimate Resource*. Princeton, N.J.: Princeton University Press, 1981.

Westing, Arthur H., ed. *Global Resources and International Conflict*. New York: Oxford University Press, 1986.

Bibliographical Essay

The literature analyzing conceptual approaches to global resource issues is indeed voluminous. The most important works include Ruth Arad and others, eds., *Sharing Global Resources* (New York: McGraw-Hill, 1979); Richard Barnet, *The Lean Years: Politics in the Age of Scarcity* (New York: Simon and Schuster, 1980); Lester R. Brown and others, eds., *State of the World 1987* (New York: W. W. Norton, 1987); Emery N. Castle and Kent A. Price, eds., *U.S. Interests and Global Natural Resources: Energy, Minerals, Food* (Washington, D.C.: Resources for the Future, 1983); Kenneth A. Dahlberg and others, *Environment and the Global Arena* (Durham, N.C.: Duke University Press, 1985); Ted Robert Gurr, "On the Political Consequences of Scarcity and Economic Decline," *International Studies Quarterly*, 29 (March 1985), 51–75; John Kincaid, "Of Time, Body, and Scarcity: Policy Options and Theoretic Considerations," *International Political Science Review*, 4 (1983), 401–16; William Ophuls, *Ecology and the Politics of Scarcity* (San Francisco: W. H. Freeman, 1977); Dennis Pirages, *Global Ecopolitics* (North Scituate, Mass.: Duxbury Press, 1978); and Julian L. Simon, *The Ultimate Resource* (Princeton, N.J.: Princeton University Press, 1981). Of these, only Simon takes a highly skeptical stance about the severity of global resource scarcity.

A far smaller number of works deals specifically with conflict over the world's resources. The most important are Nazli Choucri and Robert C. North, *Nations in Conflict: National Growth and International Violence* (San Francisco: W. H. Freeman, 1975); Philip Connelly and Robert Perlman, *The Politics of Scarcity: Resource Conflicts in International Relations* (London: Oxford University Press, 1975); Johan Galtung, *Environment, Development and Military Activity* (Oslo: Universitetsforlaget, 1982); Geoffrey Kemp, "Scarcity and Strategy," *Foreign Affairs*, 56 (January 1978): 396–414; Klaus Knorr and Frank N. Trager, eds., *Economic Issues and National Security* (Lawrence, Kan.: Regents Press, 1977); Robert C. North, "Toward a Framework for the Analysis of Scarcity and Conflict," *International Studies Quarterly*, 21 (December 1977), 569–91; Robert L. Paarlberg, "Food, Oil, and Coercive Resource Power," *International Security*, 3 (Fall 1978), 3–19; and Arthur H. Westing, ed., *Global Resources and International Conflict* (New York: Oxford University Press, 1986). Some of these writings, such as those by Kemp and Knorr and Trager, emphasize national security issues; while others, such as those by North and Westing, deal with a broad environmental approach to conflict.

More specifically, the major studies of the whaling confrontation are Center for Action on Endangered Species, *The Whalebook* (Washington, D.C.: Center for Environmental Education, 1978); David Day, *The Whale War* (San Francisco: Sierra Club Books, 1987); Marc Leepson, "Whaling: End of an Era," *Editorial Research Reports*, 2 (September 27, 1985), 718–736; Robert Mandel, "Transnational Resource Conflict: The Politics of Whaling," *International Studies Quarterly*, 24 (March 1980), 99–127; and John R. Schmidhauser and George O. Totten III, eds., *The Whaling Issue in U.S.-Japan Relations* (Boulder, Colo.: Westview Press, 1978). The Schmidhauser and Totten book is alone in avoiding a pro-conservationist, anti-whaling viewpoint.

Turning to the oil crisis, the vast literature is well represented by a special issue of *Daedalus* (vol. 104, Fall 1975) devoted entirely to the topic; Roy E. Licklider, "The Failure of the Arab Oil Weapon in 1973–1974," *Comparative Strategy*, 3 (1982), 365–380; Peter R. Odell, *Oil and World Politics*, 5th ed. (New York: Penguin Books, 1979); Anthony Sampson, *The Seven Sisters: The Great Oil Companies and the World They Shaped* (New York: Bantam Books, 1976); and Benjamin Shwadran, *Middle East Oil Crises Since 1973* (Boulder, Colo.: Westview Press, 1986). Shwadran presents the most unorthodox interpretation of the origins of the 1973–74 Arab oil embargo.

The American-Soviet episodes of grain coercion are best presented by Nick Butler, "The US Grain Weapon: Could it Boomerang?" *The World Today*, 39 (1983), 52–59; Raymond F. Hopkins and Donald J. Puchala, *Global Food Interdependence: Challenge to American Foreign Policy* (New York: Columbia University Press, 1980); Dan Morgan, *Merchants of Grain* (New York: Penguin Books, 1980); Robert L. Paarlberg, "Lessons of the Grain Embargo," *Foreign Affairs*, 59 (1980), 144–162; and Peter Wallensteen, "Scarce Goods as Political Weapons: The Case of Food," *Journal of Peace Research*, 13 (1976), 277–298. The Hopkins and Puchala and Wallensteen writings place the grain issue in the more general context of international conflict over food, while Butler, Morgan, and Paarlberg focus more on the details of recent grain confrontations.

Moving to the strategic minerals threat, the burgeoning work on the subject is exemplified by James T. Bennett and Walter E. Williams, *Strategic Minerals: The Economic Impact of Supply Disruptions* (Washington, D.C.: Heritage Foundation, 1981); David G. Haglund, "The New Geopolitics of Minerals: An Inquiry into the Changing International Significance of Strategic Minerals," *Political Geography Quarterly*, 5 (1986), 221–240; Uri Ra'anan and Charles M. Perry, eds., *Strategic Minerals and International Security* (Washington, D.C.: Pergamon, 1985); Michael Shafer, "Mineral Myths," *Foreign Policy*, 47 (1982), 154–171; and Bohdan O. Szuprowicz, *How to Avoid Strategic Minerals Shortages* (New York: John Wiley and Sons, 1981). Only Shafer seriously questions the prevailing concern about the dangers of Soviet strategic minerals actions.

Finally, the Chernobyl nuclear disaster receives its most multi-faceted coverage from a special issue (vol. 42, August/September 1986) of the *Bulletin of the Atomic Scientists* devoted to the topic; Sharon Begley and others, "The Lessons of Chernobyl," *Newsweek*, 109 (April 27, 1987), 56–57; Christopher Flavin, "Reassessing Nuclear Power," in Lester R. Brown and others, eds., *State of the World 1987* (New York: W. W. Norton, 1987), 57–80; and Bennett Ramberg, "Lessons from Chernobyl," *Foreign Affairs*, 65 (1986/87), 304–28. Studies of this incident almost without exception reflect an implicit or explicit antinuclear slant.

Index

About the Author

ROBERT MANDEL is Professor of International Affairs at Lewis and Clark College. His publications include *Perception, Decision Making, and Conflict* and *Irrationality in International Confrontation* (Greenwood Press, 1987), as well as journal articles on world politics, conflict resolution, and related areas.